Microsoft Azure Data Solutions - An Introduction

Daniel A. Seara
Francesco Milano
Danilo Dominici

Microsoft Azure Data Solutions - An Introduction

Published with the authorization of Microsoft Corporation by:
Pearson Education, Inc.

ISBN-13: 978-0-13-725250-3
ISBN-10: 0-13-725250-1

Library of Congress Control Number: 2021939201

1 2021

TRADEMARKS

Microsoft and the trademarks listed at http://www.microsoft.com on the "Trademarks" webpage are trademarks of the Microsoft group of companies. All other marks are property of their respective owners.

WARNING AND DISCLAIMER

Every effort has been made to make this book as complete and as accurate as possible, but no warranty or fitness is implied. The information provided is on an "as is" basis. The author, the publisher, and Microsoft Corporation shall have neither liability nor responsibility to any person or entity with respect to any loss or damages arising from the information contained in this book or from the use of the programs accompanying it.

SPECIAL SALES

For information about buying this title in bulk quantities, or for special sales opportunities (which may include electronic versions; custom cover designs; and content particular to your business, training goals, marketing focus, or branding interests), please contact our corporate sales department at corpsales@pearsoned.com or (800) 382-3419.

For government sales inquiries, please contact governmentsales@pearsoned.com.

For questions about sales outside the U.S., please contact intlcs@pearson.com.

EDITOR-IN-CHIEF
Brett Bartow

EXECUTIVE EDITOR
Loretta Yates

SPONSORING EDITOR
Charvi Arora

DEVELOPMENT EDITOR
Songlin Qiu

MANAGING EDITOR
Sandra Schroeder

SENIOR PROJECT EDITOR
Tracey Croom

COPY EDITOR
Kate Shoup, Polymath Publishing

INDEXER
Ken Johnson

PROOFREADER
Donna Mulder

TECHNICAL EDITOR
Joel Sjöö

EDITORIAL ASSISTANT
Cindy Teeters

COVER DESIGNER
Twist Creative, Seattle

COMPOSITOR
codeMantra

GRAPHICS
Vived Graphics

Contents

Acknowledgments   *vi*
About the Authors   *vii*
Introduction   *ix*

Chapter 1   **Understand Azure data solutions**   **1**

Data-storage concepts 2
    Types of data   2
    Understand data storage   4
    Data storage in Azure   8

Data-processing concepts 18
    Batch processing   19
    Stream processing   19
    Lambda and kappa architectures   20
    Azure technologies used for data processing   22

Use cases 26
    Advanced analytics   26
    Hybrid ETL with existing on-premises SSIS
    and Azure Data Factory   26
    Internet of things architecture   28

Summary 29

Summary exercise 31

Summary exercise answers... 31

Chapter 2   **Implement data-storage solutions**   **33**

Implement non-relational data stores 33
    Implement a solution that uses Cosmos DB,
    Azure Data Lake Storage Gen2, or Blob storage   34
    Implement partitions   46
    Implement a consistency model in Cosmos DB   48
    Provision a non-relational data store   50
    Provision an Azure Synapse Analytics workspace   70

Provide access to data to meet security requirements 74

Implement for high availability, disaster recovery,
and global distribution 82

Implement relational data stores 88

Provide access to data to meet security requirements 88

Implement for high availability and disaster recovery 97

Implement data distribution and partitions for
Azure Synapse Analytics 99

Implement PolyBase 102

Manage data security 104

Implement dynamic data masking 104

Encrypt data at rest and in motion 107

Summary 110

Summary exercise 111

Summary exercise answers 114

Chapter 3   **Manage and develop data processing
for Azure Data Solutions   115**

Batch data processing 116

Develop batch-processing solutions using
Azure Data Factory and Azure Databricks 119

Implement the Integration Runtime for Azure Data Factory 136

Create pipelines, activities, linked services, and datasets 142

Create and schedule triggers 157

Implement Azure Databricks clusters, notebooks, jobs,
and autoscaling 162

Ingest data into Azure Databricks 168

Ingest and process data using Azure Synapse Analytics 176

Streaming data 189

Stream-transport and processing engines 191

Implement event processing using Stream Analytics 192

Configure input and output 195

Select the appropriate built-in functions 201

Summary 213

Summary exercise 216

Summary exercise answers 218

Chapter 4   Monitor and optimize data solutions   219

Monitor data storage 220

    Monitor an Azure SQL Database   220

    Monitor Azure SQL Database using DMV   225

    Monitor Blob storage   226

    Implement Azure Data Lake Storage monitoring   229

    Implement Azure Synapse Analytics monitoring   229

    Implement Cosmos DB monitoring   231

    Configure Azure Monitor alerts   232

    Audit with Azure Log Analytics   236

Monitor data processing 244

    Monitor Azure Data Factory pipelines   244

    Monitor Azure Databricks   247

    Monitor Azure Stream Analytics   249

    Monitor Azure Synapse Analytics   250

    Configure Azure Monitor alerts   255

    Audit with Azure Log Analytics   256

Optimize Azure data solutions 260

    Troubleshoot data-partitioning bottlenecks   260

    Partitioning considerations   261

    Partition Azure SQL Database   263

    Partition Azure Blob storage   265

    Partition Cosmos DB   266

    Optimize Azure Data Lake Storage Gen2   267

    Optimize Azure Stream Analytics   268

    Optimize Azure Synapse Analytics   269

    Manage the data life cycle   272

Summary 276

Summary exercise 278

Summary exercise answers 279

Index   *281*

Acknowledgments

I would like to thank the following people, who helped me during the work on this book, and in my life, both professional and personal.

First, thank you to my wife, Nilda Beatriz Díaz, for helping me daily be a better person and a better professional, and for sharing with me the adventure of this life and this astounding work, all around the world.

I would also like to thank all the members of our team at Lucient, who walk with me in the path of knowledge and in the process of providing our customers the services they deserve. For this book, one team member, Mr. Joel Sjöö, was a particular help, reviewing all our technical content. Thank you, my friend.

And finally, I would like to thank Lilach Ben-Gan, who makes my English writing more readable and clearer for you, the reader, and keeps our writing work flowing smoothly and on time.

Daniel Seara

While I am used to preparing and delivering live sessions, courses, and short articles, this was my first time writing a technical book. It is a very intensive and unique experience and, at the same time, the perfect occasion to rearrange and extend my knowledge about the topics covered. But also, it is something I could not have achieved alone.

I have to say a big thank you to my wife and daughters for living many hours with a "ghost" in their house. It must not have been easy at times, but they willingly managed to give me all the time I needed.

I would also like to thank you everyone at Lucient, in particular the Italian team, which handled additional work to compensate for my months-long disappearance. Two special mentions: one is for Lilach Ben-Gan, who had the thankless task of making my English better and always understandable, and the other one is for Joel Sjöö, for making sure that the book's content adheres to the highest quality standards.

Finally, a big hug goes to my parents and parents-in-law for being our great helping hand. I really appreciate all your unrelenting efforts, and knowing you were there made the writing of this book more feasible.

Francesco Milano

I would like to thank my wonderful family, Simona, Michael, and Alice, for all their sacrifice, patience, and understanding during the countless nights and weekends without me, over the years and while I was writing this book.

I would also like to thank all my friends and colleagues at Lucient, especially Lilach Ben-Gan, for helping me to keep things on track and correcting my English writing, and Joel Sjöö, for his technical reviewing. I appreciated it a lot.

Danilo Dominici

The authors would also like to thank the team at Pearson who helped with the production of this book: Loretta Yates, Charvi Arora, and Songlin Qiu.

About the Authors

Daniel A. Seara is an experienced software developer. He has more than 20 years as a technical instructor, developer, and development consultant.

Daniel has worked as a software consultant in a wide range of companies in Argentina, Spain, and Peru. He has been asked by Peruvian Microsoft Consulting Services to help several companies in their migration path to .NET development.

Daniel was Argentina's Microsoft regional director for four years and was the first nominated global regional director, a position he held for two years. He also was the manager of the Desarrollador Cinco Estrellas I (Five Star Developer) program, one of the most successful training projects in Latin America. Daniel held Visual Basic MVP status for more than 10 years, as well as SharePoint Server MVP status from 2008 until 2014. Additionally, Daniel is the founder and "dean" of Universidad.NET, the most-visited Spanish-language site to learn .NET.

In 2005, he joined Lucient, the leading global company on the Microsoft Data Platform, where he has been working as a trainer, consultant, and mentor.

Francesco Milano has been working with Microsoft technologies since 2000.

Francesco specializes in the .NET Framework and SQL Server platform, and he focuses primarily on back-end development, integration solutions, and relational model design and implementation. He is a regular speaker at top Italian data platform conferences and workshops.

Since 2013, Francesco has also explored emerging trends and technologies pertaining to big data and advanced analytics, consolidating his knowledge of products like HDInsight, Databricks, Azure Data Factory, and Azure Synapse Analytics.

In 2015, Francesco joined Lucient, the leading global company on the Microsoft Data Platform, where he has been working as a trainer, consultant, and mentor.

Danilo Dominici is an independent consultant, trainer, and speaker, with more than 20 years of experience with relational databases and software development in both Windows and *nix environments. He has specialized in SQL Server since 1997, helping customers design, implement, migrate, tune, optimize, and build HA/DR solutions for their SQL Server environments.

Danilo has been a Microsoft Certified Trainer since 2000. He works as a trainer for the largest Microsoft learning partners in Italy, teaching SQL Server and Windows server courses. He is also a regular contributor and speaker at community events in Italy and is the co-leader of PASS Global Italian Virtual Chapter, the Italian-speaking virtual group. For his commitment, Microsoft has recognized Danilo as a Data Platform MVP since 2014.

Danilo was a SQL Server and PostgreSQL DBA and a VMWare administrator in Regione Marche, a local government organization in Italy, from 2004 to 2017. Later he became a DBA and a web developer within the Polytechnic University of Marche.

Danilo has been part of Lucient since 2011.

Introduction

Welcome to *Microsoft Azure Data Solutions - An Introduction*, a book developed by the team at Lucient Data. Cloud technologies are advancing at an accelerating pace, supplanting traditional relational and data warehouse storage solutions with new and novel non-relational options. This book explains the ins and outs of working with relational, non-relational, and data warehouse solutions in the Azure platform with a focus on providing the knowledge needed to make the right decisions for implementation in your organization. The book also contains step-by-step tutorials and exercises, enabling the reader to practice the concepts presented.

Who is this book for?

Microsoft Azure Data Solutions - An Introduction is for data engineers, systems engineers, IT managers, developers, database administrators, and cloud architects. It is intended for those who have very little or no knowledge about Azure offerings and architecture and are looking to gain confidence with the Azure platform and its various tools and services for data analysis. Consolidated experience about on-premises environments and technologies can be helpful but is not required.

How is this book organized?

This book is organized into four chapters:

- Chapter 1: Understand Azure data solutions
- Chapter 2: Implement data-storage solutions
- Chapter 3: Manage and develop data processing for Azure data solutions
- Chapter 4: Monitor and optimize data solutions

The first chapter of this book explains concepts of data storage and data processing. The second chapter deals with the implementation of relational and non-relational data stores, as well as data security management. The third chapter focuses on the development of batch processing and streaming solutions. And the last chapter covers monitoring of data storage and data processing optimization of Azure data solutions.

System requirements

This book is designed to be used with an Azure account in which different services will be implemented.

The following list details the minimum system requirements needed to run the content in the book's companion website:

- Windows 8.1
- Display monitor capable of 1024 x 768 resolution
- Microsoft mouse or compatible pointing device

About the companion content

The companion content for this book can be downloaded from the following page:

MicrosoftPressStore.com/AzureDataSolutions/downloads

The companion content includes source code of samples for use with the tutorials and exercises provided in the book.

Errata, updates, and book support

We've made every effort to ensure the accuracy of this book and its companion content. You can access updates to this book — in the form of a list of submitted errata and their related corrections — at:

MicrosoftPressStore.com/AzureDataSolutions/errata

If you discover an error that is not already listed, please submit it to us at the same page.

For additional book support and information, please visit

MicrosoftPressStore.com/Support

Please note that product support for Microsoft software and hardware is not offered through the previous addresses. For help with Microsoft software or hardware, go to http://support.microsoft.com.

Stay in touch

Let's keep the conversation going! We're on Twitter here: *http://twitter.com/MicrosoftPress*.

## Chapter 1

Understand Azure data solutions

Data engineers are responsible for data-related implementation tasks. These include the following:

- Provisioning data-storage services
- Ingesting streaming and batch data
- Transforming data
- Implementing security requirements
- Implementing data-retention policies
- Identifying performance bottlenecks
- Accessing external data sources

Mastering data engineering requires learning a specific set of tools, architectures, and platforms. Where the SQL professional, for example, might master and use T-SQL as the primary data-manipulation tool, the data engineer may use additional technologies (like HDInsight and Cosmos DB) and languages (such as Python) to manipulate data in big data systems.

Additionally, moving from on-premises to cloud-based infrastructure requires some changes in the way a data engineer uses and manages data. For example, data extraction — a process by which a data engineer retrieves raw data from a structured or unstructured data pool and migrates it to a staging data repository — changes significantly when data processing is done in the cloud. With the traditional extract, transform, and load (ETL) approach, the transformation stage can be time-consuming and potentially tie up source system resources while the process completes. In the cloud, this approach can be changed to extract, load, and transform (ELT), in which data is immediately extracted and loaded into a large data repository such as Azure Cosmos DB or Azure Data Lake Storage to minimize resource contention.

Topics covered in this chapter:

- Data-storage concepts
- Data-processing concepts
- Use cases

Data-storage concepts

The amount of data generated by systems and devices has increased significantly over the last
decade. Almost everything we do digitally is recorded, generating an enormous mass of data.
New technologies, roles, and approaches to working with this data are constantly emerging. As
a data engineer, one of your main tasks is to select the appropriate storage technology to save
and recall data quickly, efficiently, and safely.

Types of data

Before analyzing which data-storage technologies are available on Azure, let's identify the
three main categories of data you may have to deal with:

- Structured data
- Semi-structured data
- Unstructured data

Structured data

Structured data — often referred to as *relational data* — is data that adheres to a strict schema,
which defines the field names, data types, and relationship between objects. This is probably
the most commonly known type of data.

Structured data can be stored in database tables with rows and columns. Key columns may
be used — for example, to indicate which column makes a row unique or how the table is
related to another table. Figure 1-1 shows an example of a relational database structure.

Structured data is easy to use when you need to query or modify it because all the data
follow the same format. But once you force a consistent structure, further evolution of the data
can be more difficult because you have to change all existing data to have it conform to the
new structure.

Semi-structured data

This is data that is not organized and does not conform to a formal structure of tables, but does
have structures — such as tags or metadata — associated with it. This allows records, fields,
and hierarchies within the data.

Common examples of semi-structured data are XML, JSON, and YAML documents. Notice
that you *can* have a schema for this data; however, that schema is not strictly enforced as it
would be with a structured relational database. The idea here is that this semi-structured data
type allows data to be very flexible in terms of adding or removing fields or entire groups of
information. Figure 1-2 shows an example of a JSON (left) and XML (right) document.

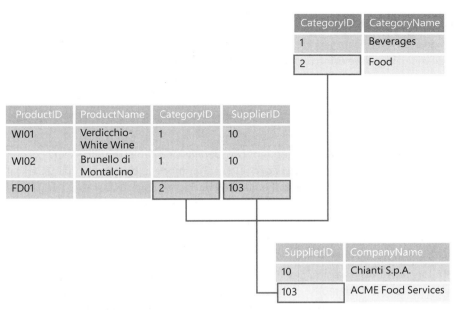

FIGURE 1-1 An example of a relational database structure

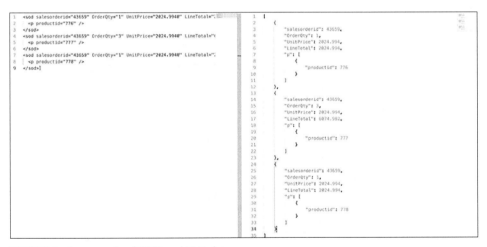

FIGURE 1-2 An example of JSON and XML data

Unstructured data

Unstructured data doesn't have any kind of structure or predefined data model, and is not organized in any particular manner that allows for traditional analysis. This is the most abundant data type we encounter.

Examples of unstructured data include PDF documents, audio or video files, JPEG images, and text files. All of these may have an internal structure that defines how they are organized internally, but this organization is unknown to the user or to the application.

Understand data storage

Depending on the type of data you are managing, the type of data storage required may differ. For example, if you are using relational data, the most appropriate data storage is a relational database. In contrast, NoSQL data can be stored in a NoSQL database such as Azure Cosmos DB or in Azure Blob storage for documents, images, or other unstructured data files.

Relational databases

Relational databases organize data as a series of two-dimensional tables with rows and columns. Every row in a table has the same set of columns. This model is based on a mathematical concept called relational theory, first described in 1969 by English computer scientist Edgar F. Codd. All data is represented in terms of tuples and grouped into relations. The purpose of the relational model is to provide a declarative method for specifying data and how to query that data. Users can directly interact with the database management system, which takes care of storing and retrieving data, to read the information they want from it.

Most vendors provide a dialect of the Structured Query Language (SQL) for retrieving and managing data. A relational database management system (RDBMS) typically implements a transactionally consistent mechanism that conforms to the atomic, consistent, isolated, durable (ACID) model for updating information. Figure 1-3 shows an example of a relational database diagram.

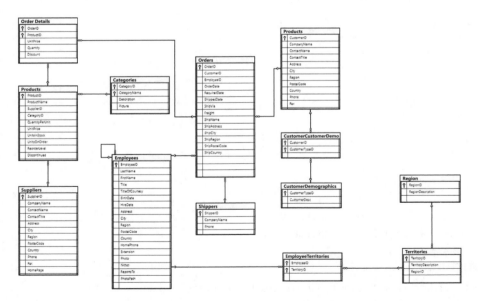

FIGURE 1-3   A diagram of a famous sample relational database called the Northwind database

An RDBMS is very useful when strong consistency guarantees are important — where all changes are atomic and transactions always leave the data in a consistent state. So, information stored in an RDBMS must be put into a relational structure by following the *normalization* process. While this process is well understood, it can lead to inefficiencies, because of the need to disassemble logical entities into rows in separate tables and then reassemble the data when running queries. Furthermore, distributing storage and processing across machines is not possible with a traditional RDBMS.

Non-relational (or NoSQL) databases

A *non-relational database* does not use the tabular schema of rows and columns found in a traditional RDBMS. Instead, these databases use a storage model that is optimized for the specific requirements of the type of data being stored.

Typical data stores in non-relational databases are key/value pairs, JSON documents, or graphs consisting of edges and vertices/nodes. What these data stores have in common is that they don't use a relational model. Instead, they tend to be more specific in the type of data they support and how that data can be queried.

DOCUMENT DATA STORE

A document data store is used to manage a set of named strings and object data values in an entity called a *document* — typically a JSON document. Each field value can be a scalar item, like a number or a string, or a compound element, such as a list or a parent-child collection.

Figure 1-4 shows an example of a JSON document with scalar values (CustomerID) and item collections (OrderItems), each one itself made by scalar values.

Key	Document
1001	```{ "CustomerID": 99, "OrderItems": [{ "ProductID": 2010, "Quantity": 2, "Cost": 520 }, { "ProductID": 4365, "Quantity": 1, "Cost": 18 }], "OrderDate": "04/01/2017" }```
1002	```{ "CustomerID": 220, "OrderItems": [{ "ProductID": 1285, "Quantity": 1, "Cost": 120 }], "OrderDate": "05/08/2017" }```

FIGURE 1-4 An example of a JSON document

This is one of the principal differences from a relational table: a JSON document can contain all the information about the entity it refers to, while a relational database must spread that data across several tables. Moreover, a JSON document does not require that all documents have the same structure, providing a high level of flexibility.

In the example shown in Figure 1-4, applications can retrieve data by using the document key, which is the unique identifier of the document. To help distribute data evenly, the document key can be hashed and can be automatically generated by the document database itself. The application can also query the documents based on the value of one or more fields. Some document databases support indexing to facilitate fast lookup of documents based on one or more indexed fields. Many document data stores support in-place updates, enabling the application to modify specific fields of the document without rewriting the entire document.

COLUMNAR DATA STORE

Where a relational database is optimized for storing rows of data, typically for transactional applications, a columnar database is optimized for fast retrieval of columns of data, typically in analytical applications. Column-oriented storage for database tables is an important factor in analytic query performance because it reduces the overall disk I/O requirements and the amount of data you need to load from disk.

Columns are divided into groups called *column families*. Each column family holds a set of columns that are logically related and are typically retrieved or manipulated as a unit. Data that is accessed separately can be stored in separate column families. This enables you to add new columns dynamically, even if rows are sparse — that is, a row doesn't need to have a value for every column. Figure 1-5 shows an example of a column family.

KEY/VALUE DATA STORE

A key/value store is essentially a large hash table. You associate each data value with a unique key, and the key/value store uses this key to store the data by using an appropriate hashing function. The hashing function is selected to provide an even distribution of hashed keys across the data storage.

Most key/value stores support only simple query, insert, and delete operations. To modify a value (either partially or completely), an application must overwrite the existing data for the entire value. In most implementations, reading or writing a single value is an atomic operation. If the value is large, writing may take some time.

An application can store arbitrary data as a set of values, although some key/value stores impose limits on the maximum size of values. The stored values are opaque to the storage system software. Any schema information must be provided and interpreted by the application. Essentially, values are blobs, and the key/value store simply retrieves or stores the value by key.

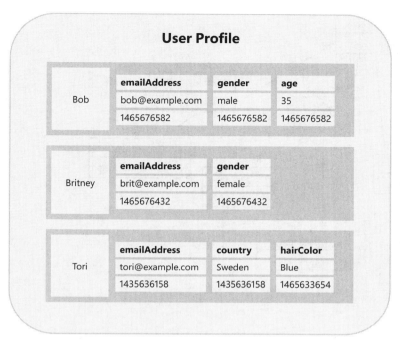

User Profile

Bob	emailAddress	gender	age
	bob@example.com	male	35
	1465676582	1465676582	1465676582

Britney	emailAddress	gender	
	brit@example.com	female	
	1465676432	1465676432	

Tori	emailAddress	country	hairColor
	tori@example.com	Sweden	Blue
	1435636158	1435636158	1465633654

FIGURE 1-5 Example of a column family

GRAPH DATA STORES

A graph data store manages two types of information:

- **Nodes** These represent entities.
- **Edges** These specify the relationships between entities.

Similar to a relational table, nodes and edges can have properties that provide information about that node or edge or a direction indicating the nature of the relationship.

The purpose of a graph data store is to enable an application to efficiently perform queries that traverse the network of nodes and edges and to analyze the relationships between entities. The diagram in Figure 1-6 shows an organization's personnel data structured as a graph. The entities are employees and departments; the edges indicate reporting relationships and the department in which the employees work. In this graph, the arrows on the edges show the direction of the relationships.

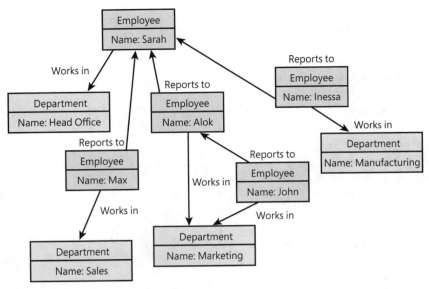

FIGURE 1-6   Example of a graph

Data storage in Azure

Now that you are more familiar with the various data types and their storage types, let's take a look at what data storage technologies are available on Azure and when to use them to properly store your data.

Azure storage account

An Azure storage account is the base storage type used within Microsoft Azure. It offers a massively scalable object store for data objects and file system services, a messaging store for reliable messaging, and a NoSQL table store.

As shown in Figure 1-7, which depicts the account type drop-down list on the Azure Portal storage account creation form, when creating a new storage account, you can choose between blob storage, which can contain only simple unstructured data such as images or videos, and two generations of Data Lake Storage.

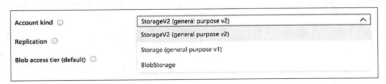

FIGURE 1-7   Azure Portal storage account creation form

Azure Blob storage

Azure Blob storage is Microsoft's object storage solution for the cloud. It is optimized for storing massive amounts of unstructured data.

Blob storage is designed for:

- Serving images or documents directly to a browser
- Storing files for distributed access
- Streaming video and audio
- Storing data for backup and restore, disaster recovery, and archiving
- Storing data for analysis by an on-premises or Azure-hosted service
- Writing to log files

Blob storage can store block blobs and append blobs into containers, which are the logical space where data is organized. Figure 1-8 shows an Azure blob storage container, named *test*, which contains a CSV file named *MSFT.csv*.

FIGURE 1-8 Azure blob storage container

Azure storage provides several options for accessing data, based on their usage patterns. These options are called *access tier*. There are three access tiers:

- **Hot access tier** This access tier is optimized for frequent access in the storage account and is the default option for new Azure Blob storage accounts.
- **Cool access tier** This tier is optimized for large amounts of data that are infrequently accessed and stored for at least 30 days. Accessing the cool access tier can be more expensive than accessing data in the hot tier.
- **Archive access tier** This tier is available only for block blobs and is optimized for data that can tolerate a high latency in retrieving data (several hours) and that can remain in this tier for at least 180 days.

Like other Azure storage technologies, Azure Blob storage employs *redundancy options*, which define where and how data is replicated. These options include the following:

- **Locally redundant storage (LRS)** Data is copied synchronously three times within the primary region.

- **Zone-redundant storage (ZRS)**   Data is copied synchronously across three Azure availability zones in the primary region.

- **Geo-redundant storage (GRS)**   Data is copied synchronously three times in the primary region, then copied to the secondary region. To access data in the secondary region, the read-access geo-redundant storage (RA-GRS) option must be enabled.

- **Geo-zone-redundant storage (GZRS)**   Data is copied synchronously across three Azure availability zones in the primary region, then copied asynchronously to the secondary region. The RA-GRS option must be enabled to access data in the secondary region.

Once data is stored into containers, users or client applications can access objects in Azure Blob storage via HTTP/HTTPS from anywhere in the world using a broad range of access methods. These include the following:

- The Azure storage REST API
- Azure PowerShell
- Azure CLI
- An Azure storage client library

There are also a lot of client libraries that developers can use to interact with Azure blob storage from within their applications using several common languages, such as the following:

- .NET
- Java
- Node.js
- Python
- Go
- PHP
- Ruby

When to use Azure Blob storage

If your business scenario is to provision a data store that will store data but will not perform analysis on it, then creating a storage account configured as a blob store is the cheapest option and works very well with images and unstructured data. A blob store enables you to use one of several tools and technologies to ingest data. These include the following:

- Azure Data Factory
- Azure Storage Explorer
- AzCopy
- PowerShell
- Visual Studio

From a security point of view, you can control who has access to the objects contained in your blob store by using access keys or shared access signatures. A shared access signature is a URI generated by Azure Portal for blob storage, which grants specific access rights to a service, container, or object. There is also a security model that uses role-based access control (RBAC), where you can assign roles to users, groups, and applications and set permissions.

It is not possible to query data contained in your blob store objects. If you need to query it, you should move it to Azure Data Lake Storage.

Azure Data Lake Storage Gen1

Azure Data Lake Storage Gen1 (ADLS Gen1) is an enterprise-wide hyper-scale repository for big data analytic workloads. ADLS Gen1 is an Apache Hadoop file system that is compatible with Hadoop Distributed File System (HDFS) and works with the Hadoop ecosystem.

ADLS Gen1 provides unlimited storage and can store any data in its native format, without requiring prior transformation. It uses Azure Active Directory for authentication and access control lists (ACLs) to manage secure access to data.

NOTE Although it is possible to use ADLS Gen1, Microsoft suggests using Gen2, which is more complete, less expensive, and offers better performance than Gen1.

Azure Data Lake Storage Gen2

Azure Data Lake Storage Gen2 (ADLS Gen2) is a set of capabilities dedicated to big data analytics built on top of Azure Blob storage. Combining the features of two existing services, Azure Blob storage and ADLS Gen1, this new generation of ADLS lets you use capabilities like file system semantics, directory and file level security, high availability/disaster recovery.

Key features of ADLS Gen2 are as follows:

- **Hierarchical namespace** Objects and files are organized into a hierarchy of directories for efficient data access and better performance. Slashes are used in the name to mimic a hierarchical directory structure.

- **Hadoop-compatible access** You can treat data as if it's stored in an HDFS. With this feature, you can store the data in one place and access it through compute technologies including Azure Databricks, Azure HDInsight, and Azure Synapse Analytics without moving the data between environments.

- **Security** ADLS Gen2 supports ACLs and Portable Operating System Interface (POSIX) permissions. You can set permissions at the directory or file level for data stored within the data lake. This security is configurable through technologies such as Apache Hive and Adobe Spark, and through utilities such as Azure Storage Explorer. All stored data is encrypted at rest using either Microsoft- or customer-managed keys.

- **Low cost** ADLS Gen2 offers low-cost storage capacity and transactions.

- **Optimized performance** The Azure Blob Filesystem (ABFS) driver is optimized specifically for big data analytics and corresponding REST APIs are surfaced through *dfs.core.windows.net*. Moreover, ADSL Gen2 organizes stored data into a hierarchy of directories and subdirectories, much like a file system, for easier navigation. As a result, data processing requires less computational resources, reducing both time and cost.

- **Data redundancy** ADLS Gen2 takes advantage of Azure Blob replication models that provide data redundancy in a single data center with locally redundant storage (LRS), or to a secondary region using geo-redundant (GRS) storage. This feature ensures that your data is always available and protected if catastrophe strikes.

Depending on the source data, you can use several tools to ingest data into ADLS Gen2, including Azure Data Factory, Apache DistCp, Apache Sqoop, Azure Storage Explorer, and AzCopy. You can also ingest events using Apache Storm. (See Figure 1-9.)

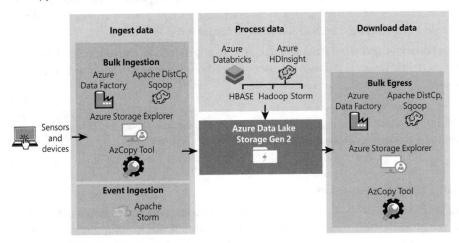

FIGURE 1-9 ADLS Gen2 ingestion tools

You can use REST APIs to interact with ADLS Gen2 programmatically. To visualize data, you can use Power BI, which can access ADLS Gen2 directly. Query acceleration is a new capability that enables applications and analytics frameworks to dramatically optimize data processing by retrieving only the data they require to perform a given operation. This reduces the time and processing power needed to gain critical insights into stored data.

When to use ADLS Gen2

ADLS Gen2 is designed for customers who need to store massive amounts of data for big data analytics use cases — for example, a medical research center that stores and analyzes petabytes of data and billions of records of related sample data as part of their research.

Azure SQL Database

Azure SQL Database is a fully managed platform as a service (PaaS) database engine designed for cloud applications when near-zero administration and enterprise-grade capabilities are needed. Azure SQL Database is always running on the last stable version of the SQL Server database engine and patched OS, with 99.999% availability. In fact, the newest capabilities of SQL Server are released first to Azure SQL Database, and then to SQL Server itself. Azure SQL Database also enables the use of advanced query processing features such as high-performance in-memory technologies and intelligent query processing.

> **NEED MORE REVIEW?** For more information about high-performance in-memory technologies, see *https://docs.microsoft.com/en-us/azure/azure-sql/in-memory-oltp-overview*. For more on intelligent query processing, see *https://docs.microsoft.com/en-us/sql/relational-databases/performance/intelligent-query-processing?toc=%2Fazure%2Fsql-database%2Ftoc.json&view=sql-server-ver15*.

There are three deployment options for Azure SQL Database:

- **Single database** Azure SQL Database is a fully managed platform as a service (PaaS) database engine that handles most database-management functions. This flavor represents a fully managed, isolated database. It is similar to the contained database in the SQL Server database engine and is perfect for cloud applications or microservices that need a single data source.

- **Elastic pool** This is a collection of single databases that have varying and unpredictable usage demands. This deployment option can be used to lower costs when managing multiple databases by sharing resources like CPU and memory at a set price.

- **Managed Instance** This is a fully managed database that has near 100% compatibility with the latest SQL Server Enterprise Edition database engine. It allows existing SQL Server customers to lift and shift their on-premises applications to the cloud with minimal application and database changes, while preserving all PaaS capabilities. These include automatic patching and version updates, automated backups, and high availability, which drastically reduce management overhead.

Azure SQL Database (except Managed Instances) can be purchased as two different models:

- **Database transaction unit (DTU)** A DTU represents a blended measure of CPU, memory, reads, and writes. The DTU-based purchasing model offers a set of preconfigured bundles of compute resources and storage to drive different levels of application performance. Having a fixed monthly payment for the selected bundle allows for a simpler way to choose from among the various offers. In the DTU-based purchasing model, you can choose between the basic, standard, and premium service tiers for Azure SQL Database, each with different resources limitations. The DTU-based purchasing model isn't available for Azure SQL Managed Instance.

- **vCore**   Compared to the DTU-based purchasing model, this purchasing model provides several advantages, including higher compute, memory, I/O, and storage limits, as well as greater availability of hardware generation models that better match the resource requirements of the workload. This model enables you to focus on critical domain-specific database administration and optimization activities.

When to use Azure SQL Database

Azure SQL Database is the right storage solution for applications or services that need a relational database in the cloud.

Azure SQL Managed Instance

Azure SQL Managed Instance (SQL MI) is a cloud database service that combines the broadest SQL Server database engine compatibility with all the benefits of a fully managed PaaS model.

SQL MI has close to 100% compatibility with the latest SQL Server (Enterprise Edition) database engine, providing a native Azure Virtual Network (VNet) implementation that addresses common security concerns. At the same time, SQL MI preserves all PaaS capabilities — such as automatic patching and version updates, automated backups, and high availability — that drastically reduce management overhead and total cost of ownership (TCO).

NEED MORE REVIEW?   For more information about automated backups, see *https://docs. microsoft.com/en-us/azure/azure-sql/database/automated-backups-overview?tabs=single-database*. For more on high availability, see *https://docs.microsoft.com/en-us/azure/azure-sql/ database/high-availability-sla*.

When to use Azure SQL MI

Azure SQL MI is best suited for customers who need to lift and shift their on-premises applications to the cloud with minimal application and database changes.

Azure Synapse Analytics

Azure Synapse Analytics — formerly known as Azure SQL Data Warehouse — is an analytics service that brings together enterprise data warehousing and big data analytics. The latest version combines these two realms with a unified experience to ingest, prepare, manage, and serve data for immediate business intelligence (BI) and machine learning (ML) needs.

Azure Synapse Analytics consists of four components:

- **Synapse SQL**   This is a complete analytics service based on the T-SQL language and supports SQL pools and SQL on-demand.
- **Apache Spark**   This is an open-source unified analytics engine for large-scale data processing.
- **Synapse pipelines**   These allow for hybrid data integration.
- **Synapse Studio**   This is the core tool used to administer and operate different features of Azure SQL Analytics.

Synapse SQL leverages a scale-out architecture to distribute computational processing of data across multiple nodes. The unit of scale is an abstraction of compute power called a data warehouse unit, which is calculated using CPU, memory, and I/O values. Compute is separate from storage, which enables you to scale compute independently of the data in your system.

Synapse offers various types of pools, which you can create within a workspace to manage data:

- **Dedicated SQL pool (formerly SQL DW)**   A collection of analytic resources provisioned when you use Azure Synapse SQL. The size of a dedicated SQL pool is predetermined by the assignment of Data Warehouse Units (DWUs).
- **Serverless SQL pool**   A serverless query service that enables you to run SQL queries on files placed in Azure storage.
- **Serverless Spark pool**   An in-memory distributed processing framework for big data analytics.

Figure 1-10 shows a logical architecture diagram of Synapse SQL pool. It involves the following:

- **Massively parallel processing (MPP)**   This is the core operation that underlies the pool. MPP uses many CPUs to run the workload in parallel when executing queries. Each CPU has its own memory and disk.
- **Control nodes**   These are the worker bees in a Synapse SQL pool. They receive queries from applications and maintain communications among all the nodes.
- **Data Movement Service (DMS)**   This moves data around the nodes whenever needed and enables parallelism in operations among compute nodes like queries or data loading.

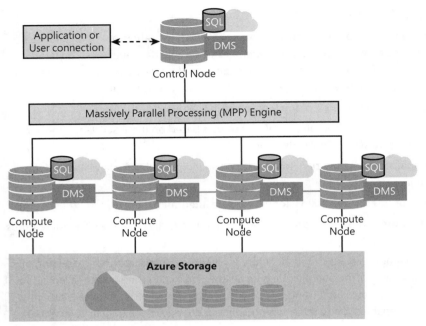

FIGURE 1-10 Azure Synapse Analytics pool architecture

> **NOTE** With the rebranding of Azure SQL Data Warehouse as Azure Synapse Analytics, Microsoft not only changed the name, but it added a lot of new functionalities to build a bridge between big data and data warehousing technologies.

> **NOTE** Azure Synapse Analytics features are comparable to Azure Databricks. For a brief comparison of the two, see the sidebar titled "When to Use Azure Databricks Versus Azure Synapse Analytics" in the "Azure Databricks" section later in this chapter.

Azure Cosmos DB

Azure Cosmos DB is Microsoft's globally distributed, multi-model database service. It is built around some key concepts:

- It must be highly responsive.
- It must be always online.
- It must be able to elastically and independently scale throughput and storage across any number of Azure regions worldwide.

Figure 1-11 provides an overview of Azure Cosmos DB features.

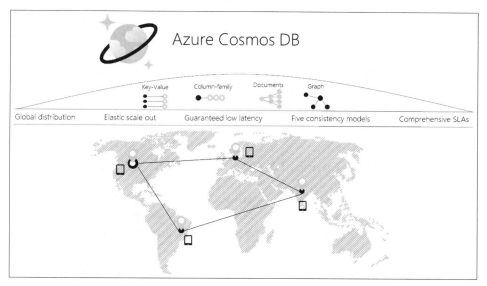

FIGURE 1-11 Azure Cosmos DB features overview

One key concept in Cosmos DB is data distribution. You can configure Cosmos DB databases to be globally distributed and available in any of the Azure regions. Putting the applications in the same region enables you to achieve minimum latency and maximum speed to access data wherever it is. Which region you choose depends on the global reach of your application and where your users are located. Cosmos DB transparently replicates the data to all the regions associated with your Cosmos account. It provides a single system image of your globally distributed Azure Cosmos database and containers that your application can read and write to locally.

With Azure Cosmos DB, you can add or remove regions associated with your account at any time. The application need not be paused or redeployed to add or remove a region. It continues to be highly available at all times because of the multi-homing capabilities that the service natively provides.

Azure Cosmos DB can be deployed using several API models, including the following:

- SQL API
- MongoDB API
- Cassandra DB API
- Gremlin DB API
- Table API

This multi-model architecture enables the database engineer to leverage the inherent capabilities of each model such as MongoDB for semi-structured data, Cassandra for columnar storage, or Gremlin for graph databases. Using Gremlin, for example, the data engineer could create graph entities and perform graph query operations to perform traversals across vertices and edges, achieving sub-second response time for complex scenarios like natural language processing (NLP) or social networking associations.

Applications built on SQL, MongoDB, or Cassandra will continue to operate without, or with minimal, changes to the application despite the database server being moved from SQL, MongoDB, or Cassandra to Azure Cosmos DB.

Azure Cosmos DB also guarantees unparalleled speed, with less than 10 ms latency for reads (indexed) and writes in every Azure region worldwide. In addition, multi-region writes and data distribution to any Azure region simplify the creation of worldwide cloud applications that need to access data within the boundary of the region to which they are deployed.

When to use Azure Cosmos DB

Data engineers should deploy Azure Cosmos DB when a NoSQL database of the supported API model is needed on a global scale and when low-latency performance is required. Azure Cosmos DB supports five 9s uptime (99.999%) and can support sub-10ms response times when provisioned correctly.

As an example, an eCommerce retailer based in London, UK, can improve its Australian customers' satisfaction by implementing a solution that provides a local version of the data in that region, using the Microsoft Australia East data center. Migrating the existing on-premises SQL Database to Cosmos DB using the SQL API would improve performance for Australian customers, as the data can be stored in the UK and replicated to Australia to improve throughput times.

Data-processing concepts

Data processing is the conversion of raw data to meaningful information. Depending on the type of data you have and how it is ingested into your system, you can process each data item as it arrives (stream processing) or store it in a buffer for later processing in groups (batch processing).

Data processing plays a key role in big data architectures. It's how raw data is converted into actionable information and delivered to businesses through reports or dashboards. (See Figure 1-12.)

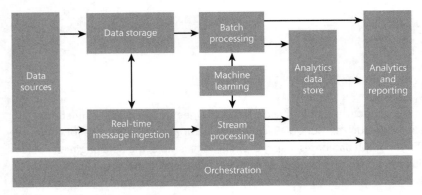

FIGURE 1-12 Big data processing diagram

Batch processing

Batch processing is the processing of a large volume of data all at once. This data might consist of millions of records per day and can be stored in a variety of ways (file, record, and so on). Batch-processing jobs are typically completed simultaneously in non-stop, sequential order. Batch processing might be used to process all the transactions a financial firm submits over the course of a week. It can also be used in payroll processes, line-item invoices, and for supply-chain and fulfillment purposes.

Batch processing is an extremely efficient way to process large amounts of data collected over a period of time. It also helps to reduce labor-related operational costs because it does not require specialized data entry clerks to support its functioning. It can also be used offline, which gives managers complete control over when it starts—for example, overnight or at the end of a week or a pay period.

There are a few disadvantages to using batch-processing software. First, debugging these systems can be tricky. If you don't have a dedicated IT team or professional to troubleshoot, fixing the system when an error occurs could be difficult, perhaps requiring an outside consultant to assist. Second, although companies usually implement batch-processing software to save money, the software and training involves a decent amount of expense in the beginning. Managers will need to be trained to understand how to schedule a batch, what triggers them, and what certain notifications mean.

Batch processing is used in many scenarios, from simple data transformations to more complex ETL pipelines. Batch processing typically leads to further interactive data exploration and analysis, provides modeling-ready data for machine learning, or writes to a data store that is optimized for analytics and visualization. For example, batch processing might be used to transform a large set of flat, semi-structured CSV or JSON files into a structured format for use in Azure SQL Database, where you can run queries and extract information, or Azure Synapse Analytics, where the objective might be to concentrate data from multiple sources into a single data warehouse.

Stream processing

Stream processing — or real-time processing — enables you to analyze data streaming from one device to another almost instantaneously. This method of continuous computation happens as data flows through the system with no compulsory time limitations on the output. With the almost instant flow, systems do not require large amounts of data to be stored.

Stream processing is highly beneficial if the events you wish to track are happening frequently and close together in time. It is also best to use if the event needs to be detected right away and responded to quickly. Stream processing, then, is useful for tasks like fraud detection and cybersecurity. If transaction data is stream-processed, fraudulent transactions can be identified and stopped before they are even complete. Other use cases include log monitoring and analyzing customer behavior.

One of the biggest challenges with stream processing is that the system's long-term data output rate must be just as fast or faster than the long-term data input rate. Otherwise the system will begin to have issues with storage and memory. Another challenge is trying to figure out the best way to cope with the huge amount of data that is being generated and moved. To keep the flow of data through the system moving at the most optimal level, organizations must create a plan to reduce the number of copies, to target compute kernels, and to use the cache hierarchy in the best way possible.

Lambda and kappa architectures

Lambda and kappa architectures are the state-of-the-art for workload patterns for handling batch and streaming big data workloads.

Lambda architecture

The key components of the lambda architecture are as follows:

- **Hot data-processing path**   This applies to streaming-data workloads.
- **Cold data-processing path**   This applies to batch-processed data.
- **Common serving layer**   This combines outputs for both paths.

The goal of the architecture is to present a holistic view of an organization's data, both from history and in near real-time, within a combined serving layer. (See Figure 1-13.)

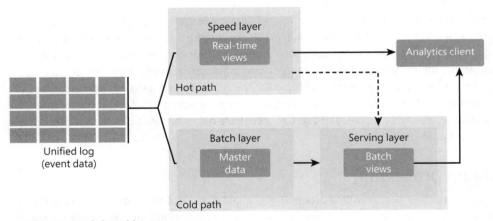

FIGURE 1-13   Lambda architecture

Data that flows into the hot path is constrained by latency requirements imposed by the speed layer, so it can be processed as quickly as possible (with some loss in terms of accuracy in favor of speed). Data flowing into the cold path, on the other hand, is not subject to the same low-latency requirements. This allows for high-accuracy computation across large data sets, which can be very time-intensive.

Eventually, the hot and cold paths converge at the analytics client application. If the client needs to display timely, yet potentially less-accurate, data in real time, it will acquire its result from the hot path. Otherwise, it will select results from the cold path to display less-timely but more-accurate data.

The raw data stored at the batch layer is immutable. Incoming data is always appended to the existing data, and the previous data is never overwritten. Any changes to the value of a particular datum are stored as a new timestamped event record. This allows for recomputation at any point in time across the history of the data collected. The ability to recompute the batch view from the original raw data is important because it allows for new views to be created as the system evolves.

Distributed systems architects and developers commonly criticize the lambda architecture due to its complexity. A common downside is that code is often replicated in multiple services. Ensuring data quality and code conformity across multiple systems — whether massively parallel processing (MPP) or symmetrically parallel system (SMP) — calls for the same best practice: reproducing code the least number of times possible. (You reproduce code in lambda because different services in MPP systems are better at different tasks.)

Kappa architecture

Kappa architecture (see Figure 1-14) proposes an immutable data stream as the primary source of record. This mitigates the need to replicate code in multiple services and solves one of the downsides of the lambda architecture.

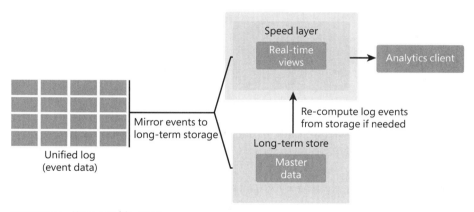

FIGURE 1-14 Kappa architecture

Kappa architecture, whose development is attributed to Jay Kreps, CEO of Confluent, Inc. and co-creator of Apache Kafka, proposes an immutable data stream as the primary source of record rather than point-in-time representations of databases or files. In other words, if a data stream containing all organizational data can be persisted indefinitely (or for as long as use cases might require), then changes to the code can be replayed for past events as needed. This allows for unit testing and revisions of streaming calculations that lambda does not support. It also eliminates the need for a batch-based ingress processing, as all data is written as events to the persisted stream.

Azure technologies used for data processing

Microsoft Azure has several technologies to help with both batch and streaming data processing. These include Azure Stream Analytics, Azure Data Factory, and Azure Databricks.

Azure Stream Analytics

Azure Stream Analytics is a real-time analytics and complex event-processing engine designed to analyze and process high volumes of fast streaming data from multiple sources simultaneously. (See Figure 1-15.)

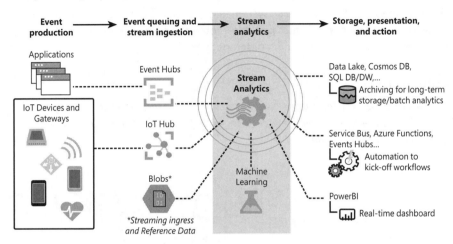

FIGURE 1-15 Azure Stream Analytics sources and destinations

Azure Stream Analytics can identify patterns and relationships among information extracted from the input sources, including sensors, devices, social media feeds, and trigger actions or initiate workflows.

Some common scenarios in which Azure Stream Analytics can be used are as follows:

- Web logs analytics
- Geospatial analytics for fleet management and driverless vehicles
- Analysis of real-time telemetry streams from IoT devices
- Remote monitoring and predictive maintenance of high-value assets
- Real-time analytics on point-of-sale data for inventory control and anomaly detection

An Azure Stream Analytics job consists of an input, a query, and an output. The engine can ingest data from Azure Event Hubs (including Azure Event Hubs from Apache Kafka), Azure IoT Hub, and Azure Blob storage. The query, which is based on SQL query language, can be used to easily filter, sort, aggregate, and join streaming data over a period of time.

Azure Data Factory

Azure Data Factory (ADF) is a cloud-based ETL and data-integration service that enables you to quickly and efficiently create automated data-driven workflows (called *pipelines*) to orchestrate data movement and transform data at scale without writing code. These pipelines can ingest data from disparate data stores. Figure 1-16 shows a simple ADF pipeline that imports data from a CSV file reading from Azure Blob storage into an Azure SQL Database.

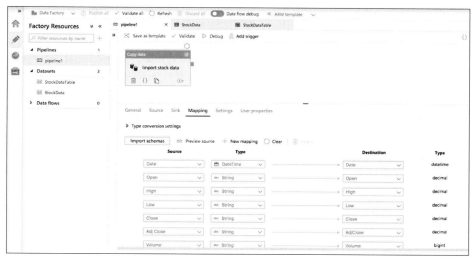

FIGURE 1-16 Sample ADF pipeline

In a pipeline, you can combine Copy Data activity — used for simple tasks such as copying from one of the many data sources to a destination (called a *sink* in ADF) — and data flows. You can perform data transformations visually or by using compute services such as Azure HDInsight, Azure Databricks, and Azure SQL Database.

Azure Data Factory includes integration runtime components, which specify where the compute infrastructure that will run an activity gets dispatched from (whether it's a public or private network). There are three types of integration runtimes:

- **Azure integration runtimes**   These can be used to copy or transform data or execute activities between cloud resources.
- **Self-hosted integration runtimes**   These are required to integrate cloud and on-premises resources.
- **Azure SQL Server Integration Services (SSIS) integration runtimes**   These can be used to execute SSIS packages through Azure Data Factory.

When to use ADF

The data movement, lineage, monitoring, and orchestration capabilities of ADF are extremely robust. You can also run external services from ADF such as HDInsight jobs. ADF automatically handles the spin-up and teardown of clusters for you. So, while you have lots of options for analytics, ML, querying, storage, and compute services, if you're considering any type of big data workload in Azure, then including ADF in your architecture will simplify and accelerate your development cycle.

Azure Databricks

Azure Databricks is a fully managed version of the popular open-source Apache Spark analytics and data-processing engine. It provides an enterprise-grade, secure, cloud-based big data and ML platform. Figure 1-17 shows the workspace starting page of a newly created Azure Databricks service, where users can manage clusters, jobs, models, or data (using Jupyter notebooks).

FIGURE 1-17   Azure Databricks workspace starting page

In a big data pipeline, raw or structured data is ingested into Azure through Azure Data Factory in batches or streamed in near real-time using Event Hub, IoT Hub, or Kafka. This data is then stored in Azure Blob storage or ADLS. As part of your analytics workflow, Azure Databricks can read data from multiple data sources — including Azure Blob storage, ADLS, Azure Cosmos DB, and Azure SQL Synapse — and turn it into breakthrough insights using Spark.

Apache Spark is a parallel processing framework that supports in-memory processing to boost the performance of big data analytic applications on massive volumes of data. This means that Apache Spark can help in several scenarios, including the following:

- **ML** Spark contains several libraries, such as MLib, that enable data scientists to perform ML activities. Typically, data scientists would also install Anaconda and use Jupyter notebooks to perform ML activities.

- **Streaming and real-time analytics** Spark has connectors to ingest data in real time from technologies like Kafka.

- **Interactive data analysis** Data engineers and business analysts can use Spark to analyze and prepare data, which can then be used by ML solutions or visualized with tools like Power BI.

Azure Databricks provides enterprise-grade security, including Azure Active Directory integration, role-based controls, and SLAs that protect stored data.

You can connect to Databricks from Excel, R, or Python to perform data analysis. There is also support for third-party BI tools, which can be accessed through JDBC/ODBC cluster endpoints such as Tableau.

When to use Azure Databricks versus Azure Synapse Analytics

Azure Synapse Analytics includes Apache Spark, just as Azure Databricks does. So, when should you use one over the other? Based on current available features, here are the recommendations for specific use cases:

- **ML development** Use Azure Databricks because it goes deeper into ML features within Spark and provides a more comfortable developer experience.

- **Ad-hoc data lake discovery** Use either Databricks or Synapse, depending on your knowledge of the tools and which GUI you prefer. (A BI developer will probably use Synapse, while a data scientist will likely use Databricks.)

- **Real-time transformations** Use Databricks to employ Spark's structured streaming and related advanced transformations and load real-time data into a Delta Lake. (A *Delta Lake* is an open-source storage layer that allows for ACID transactions the handling of scalable metadata and unifies streaming and batch data processing. Delta Lake is fully compatible with Apache Spark APIs.)

- **SQL analysis and data warehousing** Use Synapse.

- **Reporting and self-service BI** Use Synapse.

Use cases

One of the best ways to learn a new technology is to read and practice with use cases that mimic the real-world problems you have to solve and to anticipate those you might face in your business. This section analyzes some of the common scenarios for which businesses need to innovate or modernize their infrastructure and that require the skills and competencies of a data engineer to be successfully addressed.

Advanced analytics

A modern data warehouse lets you bring together all your data in a single place, at any scale, and get insights through analytical dashboards or reports. Figure 1-18 shows the architecture of this data warehouse.

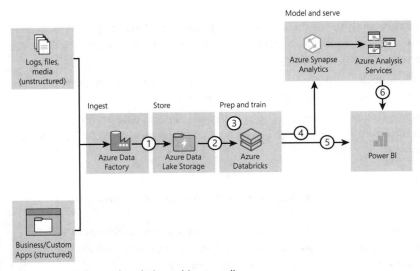

FIGURE 1-18 Advanced analytics architecture diagram

All your structured, semi-structured, and unstructured data (logs, files, and media) can be combined using Azure Data Factory and ported to an Azure Blob storage (or ADLS if you need a hierarchical storage). Then the data is cleaned and transformed using Azure Databricks and stored in Azure Synapse Analytics. From there, it can be presented to users through operational reports or analytical dashboards via Power BI.

Hybrid ETL with existing on-premises SSIS and Azure Data Factory

Many organizations are still in the middle of their digital transformation. They have on-premises data that must be combined and integrated with cloud-based data to obtain insights. Reworking existing ETL processes built with SQL Server Integration Services (SSIS) can result in a migration issue.

To facilitate a "lift-and-shift" migration of an existing SQL Server database and related SSIS packages, a hybrid approach must be used. Azure Data Factory can be employed as the primary orchestration engine, while continuing to access and execute on-premises SSIS packages via an integration runtime (IR). An IR can connect from Azure Data Factory to different on-premises resources such as SQL Server, file systems, and other data sources. It is also possible to execute local SSIS packages.

Several on-premises use cases are listed here:

- Preparing data for analytical reporting
- Loading sales data into a data warehouse for sales forecasting using ML algorithms
- Automating loading of operational data stores or data warehouses for later analysis
- Loading network router logs to a database for analysis

Figure 1-19 summarizes a sample data flow:

1. Azure Data Factory obtains data from an Azure Blob storage and executes a SSIS package hosted on-premises via IR.

2. The data-cleansing job processes both the flat files and the SQL Server database to prepare the data.

3. Data is sent back to Azure Data Factory and loaded into tables in Azure Synapse Analytics.

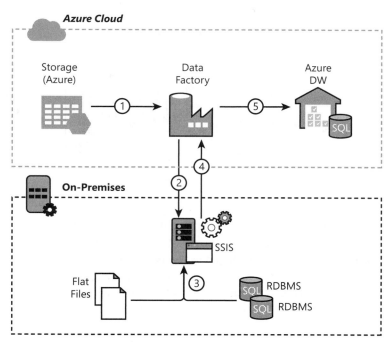

FIGURE 1-19 Hybrid ETL using ADF and SSIS

Internet of things architecture

Internet of things (IoT) applications connect devices (things) that send data from sensors or events that generates insights. Applications then analyze these insights and generate actions that improve a process or business. For example, an IoT solution might gather data from a telemetry device connected to a car's telemetry system to extract data such as the engine temperature, RPMs, and GPS information. This could facilitate the notification of the car's owner when the engine needs a checkup or send an alert in the event of an accident to the driver's insurance company to enable it to provide immediate assistance.

Figure 1-20 shows a reference architecture for IoT solutions, including not only the components required by a lambda or kappa architecture to create hot or cold paths, but also components that guarantee the solution's security, performance, and analytical capabilities.

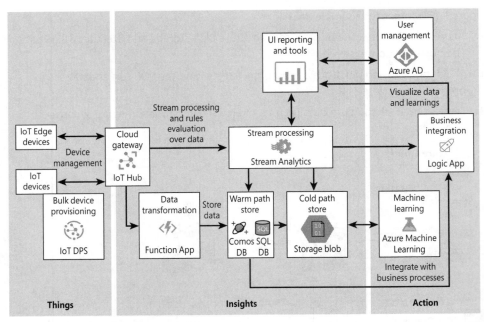

FIGURE 1-20   IoT sample reference architecture

The idea here is to ingest data as directly as possible from the physical devices or machines that generate it — for example, a temperature/pressure or counter sensor along an assembly line. IoT Hub helps the ingestion process by providing gateway services, and Azure functions transform the data as needed before it is stored in a Cosmos DB document and/or Azure SQL Database table. Azure Stream Analytics receives processed data from IoT Hub and stores it in a database or storage account, depending on the type of data. Azure Machine Learning evaluates newly added data and stores results in the cold path. Power BI or other tools can be used to extract information from the real-time data, using Active Directory credentials to secure access. To enable the customer to integrate this data with other business applications, you can extract it from multiple sources, such as Cosmos DB, Azure SQL Database, and Azure Stream Analytics, and consequently change data at the destination.

Summary

- Data engineers are responsible for data-related implementation tasks. These include provisioning data-storage services, ingesting streaming and batch data, transforming data, implementing security requirements, implementing data-retention policies, identifying performance bottlenecks, and accessing external data sources.

- There are three main types of data: structured, semi-structured, and unstructured.

- Structured data is data that adheres to a strict schema that defines field names, data types, and the relationship between objects. It is commonly known as *relational data*.

- Semi-structured data is not organized and does not conform to a formal structure of tables. It does, however, have structures associated with it, such as tags and metadata, which allow for records, fields, and hierarchies within the data. JSON and XML documents are examples of unstructured data.

- Unstructured data has no usable schema at all. Examples of unstructured data include video and image files.

- Azure SQL Database is a fully managed platform as a service (PaaS) database engine. It is designed for cloud applications when near-zero administration and enterprise-grade capabilities are needed. It always runs on the most recent stable version of the SQL Server database engine and patched OS.

- Azure SQL Database is the best solution to store relational data. It uses advanced query processing features, such as high-performance in-memory technologies and intelligent query processing.

- Azure SQL MI is a cloud database service that combines the broadest SQL Server database engine compatibility with all the benefits of a fully managed PaaS model.

- Azure Synapse Analytics, formerly known as Azure SQL Data Warehouse service, brings together enterprise data warehousing and big data analytics. It combines the power of an analysis engine with Apache Spark, an integration engine (based on Azure Data Factory), and a user interface to create and manage projects.

- NoSQL databases, used to store semi-structured data, allow for various types of data stores, including documents, columns, key-value pairs, and graphs.

- An Azure storage account is the simplest form of storage in Azure. It can be set to Blob storage, ADLS Gen1, or ADLS Gen2.

- Azure Blob storage is optimized for storing massive amounts of unstructured data.

- ADLS is designed for customers who require the ability to store massive amounts of data for big data analytics use cases.

- ADLS Gen2 is meant for big data analytics. It is built on top of Azure Blob storage.

- Azure Cosmos DB is Microsoft's globally distributed, multi-model database service. It can use several APIs to access data, including SQL, MongoDB, Cassandra, Gremlin, and Tables.

- One key concept in Cosmos DB is data distribution. You can configure Cosmos DB to replicate data geographically in real time. Applications can then access data that is stored nearest to them.

- Azure Cosmos DB guarantees less than 10 ms latency for reads (indexed) and writes in every Azure region worldwide.

- Data processing is the conversion of raw data to meaningful information. Depending on the type of data you have and how it is ingested into your system, you can process each data item as it arrives (stream processing) or store it in a buffer for later processing in groups (batch processing).

- Batch processing is the processing of a large volume of data all at once. It is an extremely efficient way to process large amounts of data collected over a period of time.

- Stream, or real-time, processing enables you to almost instantaneously analyze data streaming from one device to another.

- Lambda architecture is a specialized data-processing architecture that defines two paths where data flows from a source to a destination: a hot path and a cold path. The hot path is used for stream processing to analyze data while it flows through the pipeline. The cold path is used to store data as is for later recomputation or analysis.

- Kappa architecture enables you to build a streaming- and batch-processing system on a single technology. This means you can build a stream-processing application to handle real-time data, and if you need to modify your output, you can update your code and run it again in a batch manner.

- Azure Stream Analytics is a real-time analytics and complex event-processing engine that is designed to analyze and process high volumes of fast streaming data from multiple sources simultaneously.

- Azure Stream Analytics can ingest data from Azure Event Hubs (including Azure Event Hubs from Apache Kafka), Azure IoT Hub, and Azure Blob storage.

- Azure Data Factory (ADF) is a hybrid data-integration service that enables you to quickly and efficiently create automated data-driven workflows (called *pipelines*) without writing code.

- Pipelines created using ADF can ingest data from disparate data stores and write plain or transformed data to a destination (or *sink*).

- Azure Databricks is a fully managed version of the popular open-source Apache Spark analytics and data-processing engine. It provides an enterprise-grade and secure cloud-based big data and ML platform.

- Azure Databricks can be used with ML or real-time analysis applications because it goes deeper into ML features within Spark and provides a more comfortable developer experience.

Summary exercise

In this summary exercise, test your knowledge about the topics covered in this chapter. You can find answers to the questions in this exercise in the next section.

Northwind Electric Cars Inc. is an international car racing and manufacturing company with 500 employees. The company supports racing teams that compete in a worldwide racing series. Northwind Electric Cars has two main locations: a main office in Manchester, England, and a manufacturing plant in Modena, Italy. Most employees work in the facility in Italy.

During race weekends, Northwind Electric Cars uses an internal application called Guardian to read and analyze telemetry data sent by multiple sensors installed in the cars to the Manchester data center. This telemetry data is sent in real time to a MongoDB database and moved to a SQL Server database for a later analysis using a SQL Server Integration Services (SSIS) package.

Northwind Electric Cars is in the process of rearchitecting its data estate to be hosted in Azure, decommissioning the Manchester data center, and moving all the current applications to Azure. Technical requirements are as follows:

- Data collection from the Guardian application must be moved to Azure Cosmos DB and Azure SQL Database.
- Data must be written to the Azure data center closest to each race in the minimum amount of time.

Answer the following questions about connecting Azure services and third-party applications:

1. Which services should you use to replace MongoDB and SQL Server on-premises with an Azure solution but leave the application code as is?

2. Which technology should you implement to satisfy the requirement to write telemetry data to the closest Azure data center?

3. With what should you replace the SSIS package that copies telemetry data from MongoDB to a SQL Server data warehouse?

Summary exercise answers

This section contains the answers to the summary exercise questions in the previous section and explains why each answer is correct.

1. You should use Azure Cosmos DB with the MongoDB API to replace MongoDB, and Azure SQL Database to replace the current SQL Server implementation. Cosmos DB with MongoDB API is fully compatible with MongoDB, so you can reuse almost 100% of the application code. Azure SQL Database is a good replacement for SQL Server because, depending on the compatibility and service level required, you can choose among single, elastic pool, or managed instance flavors.

2. With Azure Cosmos DB with geo-replication, data is replicated across all the selected Azure regions, so applications can use the right database, nearest to the race location.

3. After MongoDB and SQL Server are migrated to cloud services, an Azure Data Factory pipeline can easily connect Cosmos DB and Azure SQL Database.

Chapter 2

Implement data-storage solutions

There is a lot of data out there that cannot be stored in a structural way. For example, the Word document in which this chapter was written has no common structure with other documents delivered for this book, like scripts, plan data, notebooks, and so on. So, they are impossible to store in the same table unless you consider a table with a column for the title and another for the content. This would be a very poor structure for a relational database!

This kind of content — and others like plain text, comma-separated values, Java Script Object Notation (JSON), graph structures, and binary content like audio, video, and pictures — require special platforms to store them in. In many cases, they will not be managed by relational databases. This chapter covers different methods for non-relational data storage and how to work with them.

> **NOTE** Each kind of data requires a different kind of storage. This book evaluates the various approaches to storing these different types of data. In each case, before outlining the implementation procedure, we will briefly discuss the kind of data in question and the reasons to use a specific type of storage with that data.

Topics covered in this chapter:
- Implement non-relational data stores
- Implement relational data stores
- Manage data security

Implement non-relational data stores

There are different types of non-relational data currently in use. In Azure, different services provide storage for them. Table 2-1 shows the different types of storage and the Azure services that allow you to use them. As you can see, several types of storage have more than one service supporting them, and several services could serve more than one type of storage.

TABLE 2-1 Types of storage serviced by Azure Platform

Storage	Azure Service
Key-value	Cosmos DB Azure Cache for Redis
Document	Cosmos DB
Graph	Cosmos DB
Column family	HBase in HDInsight
Search indexed information	Azure Cognitive Search
Time series	Time Series Insights
Object	Blob storage Table storage
Files	File storage

Implement a solution that uses Cosmos DB, Azure Data Lake Storage Gen2, or Blob storage

Several platforms can help you store non-relational data. In some cases, these could be very specialized platforms, like MariaDB or PostgreSQL. These are mostly for "shift-and-lift" scenarios that require you to move data from on-premises to the cloud without changing the applications. Leaving those specific cases aside, this section covers platforms that are probably more useful for distributed processing and online analytical processing (OLAP) scenarios.

Cosmos DB

If you need distributed access to data with very fast response times and high availability, Cosmos DB will probably be the appropriate storage to use.

As you saw in Table 2-1, Cosmos DB can store different kinds of data. This is because it stores each piece of data as a small atom-record-sequence (ARS). An ARS is a small chunk of one primitive data type — for example, bool, string, integer, and so on (atom) — organized as arrays (records and sequences). Any content sent to Cosmos DB is translated to ARS and stored in this format.

Having no specific schema, Cosmos DB indexes any content it receives, keeping the indexes to accelerate query responses and working as a schema-agnostic database engine. After data is queried, the results are formatted based on the query requirement — mostly in JSON format, but in other formats, too.

> **NEED MORE REVIEW?**   For more detailed information on the original work of Microsoft Research for Document DB, which is the underlying foundation of Cosmos DB, see *https://www.vldb.org/pvldb/vol8/p1668-shukla.pdf*.

Cosmos DB results in a multiengine platform. It can contain different databases that support different types of storage and serve different APIs to optimize the management of different kinds of data.

From an implementation point of view, using Cosmos DB starts with the creation of a *Cosmos DB account*. Each account can have different configurations for consistency, distribution, and so on, as you will see later in this chapter.

The account must define which API to use. The APIs supported by Cosmos DB are as follows:

- **MongoDB API**   Cosmos DB can host Mongo DB repositories, which can be imported directly into it. In fact, Cosmos DB supports importing data from several sources and can transform and import entire databases with the MongoDB format. After data is imported, you can employ the same syntax used by applications designed for MongoDB in Cosmos DB, following the JScript dotted notation for MongoDB, as in the following statement:

```
db.Items.find({},{City:1,_id:0})
```

- **Core (SQL) API**   This API is the default for Cosmos DB. It enables you to manage data the same way you do with relational storage tools. It uses a SQL-like syntax and data types from JScript, which are as follows:
  - Undefined
  - Null
  - Boolean
  - String
  - Number
  - Object

The reason for these data types is that all information is stored in JSON format and uses only standard JScript-recognized data types. You can obtain data from a sample using a SQL-like statement like the following:

```
Select City from customers
```

Core SQL defines a subset of ANSI SQL commands like SELECT, GROUP BY, ORDER BY, and WHERE, and some aggregate functions like SUM, AVG, and so on.

- **Cassandra API**   Cassandra is another type of data storage that can be imported directly to Cosmos DB. You can use this API with Cosmos DB simply by changing the connection to the new data storage. Cosmos DB supports Cassandra Query Language (CQL) version 4. For this API, the queries previously defined would be as follows:

```
Select City from customers
SELECT "City", "PostalCode" FROM Customers WHERE "Country" = 'Germany'
```

- **Gremlin API** This API processes graph data. It requires specific syntax. This information must be extracted and formatted correctly to send results to the client. Cosmos DB uses *Apache TinkerPop Gremlin* for querying. For this API, the previously defined queries — assuming the information in question is stored as graphs — would be as follows:

```
g.V().hasLabel(Customers).out ('City')
```

- **Azure Table API** When you need global replication or indexing by several properties, and the elements stored in tables are in JSON format, you can mount the data in Cosmos DB and run either or both operations using the Azure Table API. This API uses a partition ID and an entry ID to uniquely identify each entry. For example, if you need to store time series like those generated by IoT devices, you can mark the partition key as the device identity and the date and time of the data as the individual entry. Because IoT devices usually generate lots of information every second, you should use the most precise date and time format possible to avoid collisions.

To manage different APIs, you must create different accounts. This will not be a problem because you can have as many as 100 Cosmos DB accounts per subscription.

Figure 2-1 shows how storage management is designed in Cosmos DB. As you can see:

- Within the account, you define one or more databases.
- Each database has one or more containers.

- Inside each container will be items. The item is the information unit of the storage. For example, if you want to store information about countries, each country could be a JSON chunk of information, or item, with all the properties of the country inside it.
- Other elements can be stored alongside items, such as triggers, procedures, and functions.
- Outside the containers, the database stores security information, like users, permissions, and so on.

You measure resource usage in Cosmos DB database request units (RUs). An RU is an abstract representation of the CPU, input/output operations per second (IOPS), and memory usage. Because there are different types of data, it is difficult to estimate consumption. This is why the unit measures activities against the database instead of the resources themselves.

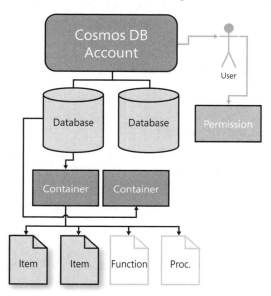

FIGURE 2-1 Cosmos DB resource model

Different factors affect RU metrics. These include the following:

- **Item size**   Larger items consume more RUs.
- **Indexing**   In Cosmos DB, everything is usually indexed. You can reduce consumption by reducing the indexing target. The number and size of the properties being indexed will affect the RUs used.
- **Consistency levels**   A high consistency level consumes more RUs than a low consistency level. (You'll learn more about consistency levels later in this chapter in the section "Implement a consistency model in Cosmos DB.")
- **Queries**   These cost more than punctual reads. Query results, types, predicates, and so on also affect RUs.
- **Procedural objects**   These include procedures, functions, and triggers.

Because there are several factors that affect RUs, you should implement good programming and auditing standards to evaluate these measurements during normal use. The Response header for each call performed on a Cosmos DB returns the RUs used.

PRACTICE   Create a Cosmos DB account

Follow these steps to create a Cosmos DB account.

> **NOTE** For the practice exercises in this chapter, you can use your own subscription. If you don't have one, refer to Chapter 1 to find out how to get a free test account.

1. Open *https://portal.azure.com* and use your credentials to log in.
2. Select your subscription.
3. Do one of the following:
   - If you have a resource group, select it and skip to step 4.
   - If you do not have a resource group, you need to create one. To do so, click the **Add**. Then, in the search box, type **Azure Cosmos DB**, and select the corresponding entry that appears in the drop-down list. Finally, click the **Create** button.
4. In the **Basics** tab, confirm your subscription and resource group, and enter the name for your account, the API you want to use, and the location.

> **NOTE** You can enable notebooks for your account. This changes the options in the Location drop-down menu. You can also select the capacity mode. At the time of this writing, the Serverless mode is in preview, so we will not use it.

5. Configure the following settings:
   - **Apply Free Tier Discount**   Select **Apply** to obtain 400 RUs and 5 GB storage free of charge. This could be applied as 25 containers for shared throughput databases.
   - **Account Type**   This setting changes the user interface experience in the portal, but not the resource utilization or functionality. Leave this set to **Non-Production**.
   - **Geo-Redundancy**   Select **Enable** to set up for geo-localization so you can define the regions you want to use for your databases.
   - **Multi Region Writes**   This option makes it so different regions can be updated. Leave this set to (**Disable**).
   - **Availability Zones**   This option enables the use of different zones in the same region for availability. Leave it set to **Disable** for now.
6. Click the **Networking** tab.

   The tab that appears enables you to select the kind of external access your account will admit, including whether you will allow other Azure resources to use the current one.
7. Enable the **External Access** option.

8. In the firewall settings, enter your IP address in the **IP Address** box if you want.

9. Click the **Backup Policy** tab and choose your backup settings, including when backups should occur and how long they should be stored.

10. Click the **Encryption** tab and specify whether the encryption process uses an auto-generated key or a custom one whose URL you specify.

11. Click the **Tags** tab and enter any custom tags you like — for example, for billing control.

12. Click the **Review and Create** tab to view the approximate time when your account will be created. (This varies depending on the region selected.) Then click the **Create** button.

> **NOTE** The deployment process may take some time to complete.

> **NOTE** The free tier discount can be applied to only one Cosmos DB account per subscription.

After your account is deployed, you add databases and containers to it to store your information. As you learned in Chapter 1, a *container* is a logical space where data is organized. Containers are an important aspect of distribution and partition implementation. (This is discussed in "Implement partitions" later in this chapter.) There is no procedure to create a database. They are created automatically when you create a container. You'll learn how to create a container in the next practice exercise.

> **NOTE** For the Cassandra API, the database is called a *keyspace*, but it is the same concept as a database. With Table API, after you create your first table, a default database is automatically created to store it.

PRACTICE Create a container

Follow these steps to create a container. (Note that we've offered some sample values, which you can use, or not, as you prefer.)

1. At the top of the **Overview** screen of your recently created Cosmos DB account, click the **Add Container** button.

2. Configure the following settings in the panel that appears to the right:

   - **Database ID** To create a new database to house the new container, click the **Create New** option. In this case, name the new database **CovidDB**.

> **NOTE** To select an existing database, click to choose the desired database.

> **NOTE** Much of this book was written during the COVID-19 pandemic, so we chose this database as an homage of sorts.

- **Provision Throughput at the Database Level** This option is selected by default. If you prefer to provision throughput at the container level, deselect it this option. Either way, you must indicate a throughput value, which is **400 RUs** by default.
- **Container ID** To assign a name to the new container, type it in this box. In this case, we've typed **GlobalData**.
- **Indexing** Leave this set to **Automatic** as is if you want the data to be automatically indexed. (This is recommended to help applications find data.)
- **Partition Key** Enter the partition key in this box to establish the values by which data will be divided in partitions. In this example, we've used **/Code** as the partition key. (We will talk about partitions in the "Implement partitions" section later in this chapter.)
- **Analytical Store** Leave this deselected for now.

Analytical Store

You enable the Analytical Store feature if you want to use the container for analytical processes. In this case, you must also enable Synapse Link for the account. Synapse Link allows for column storage in Cosmos DB, so you can use Synapse SQL or Synapse Spark to work directly with the data to perform analytical procedures in near real time without an ETL or ELT process. When you do, the Cosmos DB engine prevents impact on other OLTP workloads running simultaneously. For more detailed information about how this works, see *https://docs.microsoft.com/en-us/azure/cosmos-db/analytical-store-introduction*.

- **Primary Keys** Defining these enables applications to work when unique identifiers for each set of data are required. We will not define them at this time.

3. Click the **OK** button to create the container.

WORK WITH INFORMATION IN DATA EXPLORER

After you create the container, you will be directed to Data Explorer, where you can begin working with your information. To see your container, expand the directory tree in Data Explorer. Your container contains the following nodes and options:

- **Items**   This contains your data. Click this node to see the contents of this folder in the panel on the right. To filter these contents, click the **Change Filter** button. Use the buttons above the panel to add or upload new content.

> **NOTE**   The upload process allows up to 2 MB in the same operation. To upload more data, you must perform several uploads. Alternatively, you can execute a script or a program to upload data.

- **Settings**   Here you can change the following configuration parameters for your container:
  - **Time to Live**   The time to live (TTL) defines how long the container will exist. By default, it is null (off), which means the container will live forever. The same is true if the value here is on (no default), or –1. Other values represent the seconds from the last update date and time. The TTL can also be assigned by item (in case the TTL for the container is set to –1).
  - **Geospatial Configuration**   This specifies whether the geospatial configuration feature will relate to geography or geometry.

> **NOTE**   By default, Geospatial Configuration is set to Geography, which means the geospatial data, if any, defines geographical coordinates. If you change it to Geometry, the data will store geometric points, not map locations.

  - **Partition Key**   This is a read-only field because the partition key can be defined only when the container is created.
  - **Indexing Policy**   This enables you to modify the indexing behavior by editing a JSON configuration. The indexing mode can be Consistent or None. When the mode is Consistent, the JSON configuration can contain included and excluded paths. By default, the included path is `*` (`all content`), while the _etag attribute (which is used internally by Cosmos DB) and all its offsprings are excluded.
- **Stored Procedures**   Here you define procedures in JScript to manage and process items.
- **Functions**   These are code fragments in JScript that perform operations on items or obtain values from them and return results.
- **Triggers**   Here you can define procedures in JScript that will execute when something happens with an item — for example, when a new item is stored, deleted, and so on.

COSMOS DB PROGRAMMING

Cosmos DB uses JScript as the code language for stored procedures, functions, and triggers. As an example, to query items, you must use the partition key, and the contents are returned using the `getContext().getCollection()` JScript function.

The following code is the sample that appears by default when you create a new stored procedure:

```
// SAMPLE STORED PROCEDURE
function sample(prefix) {
    var collection = getContext().getCollection();
    // Query documents and take 1st item.
    var isAccepted = collection.queryDocuments(
        collection.getSelfLink(),
        'SELECT * FROM root r',
    function (err, feed, options) {
        if (err) throw err;
        // Check the feed and if empty, set the body to 'no docs found',
        // else take 1st element from feed
        if (!feed || !feed.length) {
            var response = getContext().getResponse();
            response.setBody('no docs found');
        }
        else {
            var response = getContext().getResponse();
            var body = { prefix: prefix, feed: feed[0] };
            response.setBody(JSON.stringify(body));
        }
    });
    if (!isAccepted) throw new Error('The query was not accepted by the server.');
}
```

Notice that, as with many JScript codes, the queryDocuments function uses another function as part of the parameters, which is responsible for managing the result.

> **NEED MORE REVIEW?**   For detailed samples of these programmatic components, follow this link: *https://docs.microsoft.com/en-us/azure/cosmos-db/how-to-write-stored-procedures-triggers-udfs*.

PRACTICE   Run queries and obtain RU information

You can run queries in Data Explorer. You can also obtain information about the RUs used in the query. To run a query, follow these steps:

1. Select the container on which you want to run the query.
2. Click the **New Query** button.
3. Type a simple test query like the one that follows and click the **Run Query** button.

    ```
    Select * from c
    ```

4. After you execute the query, click the **Query Stats** tab to see how many RUs it used. (See Figure 2-2.)

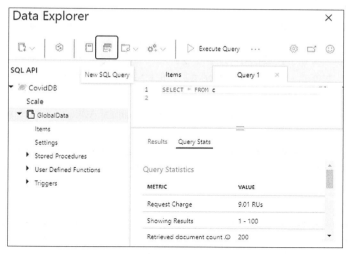

FIGURE 2-2   Viewing a query's RUs usage

5. To download detailed information about the query, like that shown in Table 2-2, click the **Per-Partition Query Metrics (CSV)** link at the bottom of the **Query Stats** tab.

TABLE 2-2   Query metrics

Property	Sample Value
Partition key range ID	0
Retrieved document count	200
Retrieved document size (in bytes)	124498
Output document count	200
Output document size (in bytes)	124798
Index hit document count	200
Index lookup time (ms)	0
Document load time (ms)	7.42
Query engine execution time (ms)	0.12
System function execution time (ms)	0
User-defined function execution time (ms)	0
Document write time (ms)	0.12

Azure Data Lake Storage Gen2 and Blob storage

If you do not need regional or geographical replication, but you do need to manage high volumes of data for data exploration, data analysis, or machine learning (ML) applications, you can use Azure Data Lake Storage (ADLS) Gen2 or Blob storage.

Usually, you use ADLS to store data in its original format, without any intermediate processes. This enables you to use the same information for different applications and analyze data using different approaches. The data remains immutable.

In addition, using ADLS does not restrict you to creating a data warehouse, because you can obtain information from the data lake and transform and store it in the needed format using ELT, as explained in Chapter 1. Moreover, ADLS can store multiple petabytes of data with hundreds of gigabits of throughput, so you can preserve all the information you want to analyze now or in the future.

You might need to use information from different analysis engines. ADLS supports the following:

- **Hadoop Distributed File System (HDFS)**   This file system supports high data throughput and intense access levels. It is tuned to handle large files based on a write-once-read-many access model. For more information about Hadoop specifications, see *https://hadoop.apache.org*.

- **Azure Blob Filesystem (ABFS) driver**   This reaches data stored in HDFS. It enables you to access data directly, without needing to use an ADLS process. This driver is used by Azure HDInsight, Azure Databricks, and Azure Synapse Analytics. It converts the object-oriented storage in blobs to a simulated file system scheme.

- **ACL and POSIX permissions**   This enables you to manage security for Windows, Unix, and Linux operating systems, with specific extensions for data lake content.

- **REST API exposure**   You can use REST API calls to access content.

> **IMPORTANT**   If you look for ADLS in the Azure Portal, you will see ADLS Gen1. We do not recommend that you use this version, as it will be deprecated sooner or later, and ADLS Gen2 has all the features discussed here.

PRACTICE   Create an Azure storage account to use as a data lake

You don't really create a data lake per se. Rather, you create an Azure storage account while enabling specific features to activate ADLS. This practice exercise shows you how.

1. Open *https://portal.azure.com* and use your credentials to log in.
2. In the search box, type **Azure storage**, and select the corresponding entry that appears in the drop-down list.
3. Click the **Add** button.
4. In the **Basics** tab, configure the following settings:
   - **Subscription**   Select your subscription.
   - **Resource Group**   Select your resource group.
   - **Storage Account Name**   Type a name for the new account.

- **Location** Choose the location. (The one associated with the resource group will be selected by default.)
- **Performance** The **Standard** option button is selected by default. Leave this as is for this practice exercise. Just be aware that selecting Premium means the storage account would exist on SSD disks, allowing faster access for reads and writes.
- **Account Kind** Select the **StorageV2 (General Purpose v2)** option from the drop-down list.
- **Replication** Because this is a practice exercise, select **Locally-Redundant Storage (LRS)** from the drop-down list to reduce costs.
- **Blob Access Tier** Leave this set to **Hot**.

5. Skip the **Networking** and **Data Protection** tabs. You'll leave the default values in those tabs as is.

6. Click the **Advanced** tab and configure the following settings:
   - **Secure Transfer Required** Set this to **Enabled**.
   - **Allow Blob Public Access** Set this to **Disabled**.
   - **Minimum TLS Version** Select **Version 1.2**.
   - **Large File Shares** Set this to **Disabled**.
   - **Hierarchical Namespace** Set this to **Enabled**. This configures the new storage account as a data lake.

7. Skip the **Tags** tab for this practice exercise.

8. Click the **Review + Create** tab. Then click the **Create** button.

There are many Azure services capable of retrieving information from Azure storage and from ADLS. Table 2-3 lists them.

TABLE 2-3   Azure services with access to ADLS

Service	Details
Azure Data Box	Migrates content from on-premises HDFS storage to ADLS
Azure Data Explorer	Enables you to manage storage content — uploading, downloading, and removing content, creating new containers, and so on

Service	Details
Azure Data Factory	A platform for managing ELT and other data-management processes (more on this in Chapter 3)
Azure Databricks	An Apache Spark–based big data analytics service designed for data science and data engineering, with full support for languages like Python, Scala, R, Java, and SQL, plus multiple data science libraries
Azure Event Grid	An event subscription service that can register listeners and spread events that happen within the Azure storage account
Azure Event Hub	Uses the storage account as a repository for event information; can receive massive entries and store them quickly in the account, to be processed later
Azure HDInsight	An open source analytic service that supports frameworks such as Apache Spark, Apache Hive, Hadoop, LLAP, Apache Kafka, Apache Storm, and R, which can use ABFS to retrieve data from ADLS
Azure IoT Hub	Obtains data from many different types of devices and stores it quickly in the ADLS, mostly as time series data
Azure Logic Apps	Coordinates the orchestration of procedures based on events and data received through multiple connectors, including Azure storage
Azure Machine Learning	A machine learning cloud service capable of obtaining information from ADLS during the modeling process
Azure Stream Analytics	A real-time analytic service that can use ADLS as a repository to store incoming information
Azure Synapse Analytics	An analysis service that obtains data from ADLS by combining SQL commands with Spark or Hive
Microsoft Power BI	A rich tool for visualizing data in several graphics formats and for analyzing and building informational dashboards that is fully capable of obtaining data from ADLS

Implement partitions

For better performance, Cosmos DB divides content into partitions. This facilitates regional and geographical replication. Although you can define logical partitions, under the hood, Cosmos DB manages them as physical partitions, which do not necessarily match up. For example, if your partitions are not very large, several logical partitions will probably reside inside a single physical partition.

Physical partitions are calculated based on various parameters, like the size of your data, and are not something you need to worry about. Logical partitions are a different story. You must choose a partition key for each container. This is the initial slice for data searching and storing. Each item will be uniquely identified by a combination of the partition key and an entry ID, which is a value that Cosmos DB assigns automatically to each item. Figure 2-3 shows how partitions might be distributed in a Cosmos DB database.

NOTE If you want to establish comparison between Cosmos DB and a relational database, you can think of the partition key as the name of the table.

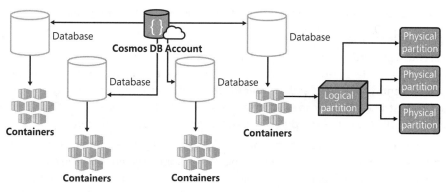

FIGURE 2-3 Cosmos DB storage

The partition key has another effect: The items in a container are isolated by partition key. This means you cannot not use the partition key as a primary key. For internal use, Cosmos DB uses the autogenerated entry ID as the primary key. However, you can define your own primary key when you define the container. Internally, Cosmos DB will manage the items by the entry ID, but you do not have to use it at all.

Choosing a partition key involves some important considerations:

- **The partition key must be an attribute (property) of each and every item** The attribute identifier must be alphanumeric, with no special symbols (except for an under-score, which is accepted).

- **The partition key value must be immutable** It will be read-only except with the insert action. If you need to change the value, you must read the item, delete it, change the value for the partition key, and insert a new one.

- **Ideally, the partition key should repeat in several items** Having many variations in the partition key value generates a large number of small partitions, which affects performance, usage, and cost.

There is a potential problem with the management of partitions. Cosmos DB supports ACID transactions for procedures, functions, and triggers, but only at the partition level.

You must rethink your storage concepts when working with Cosmos DB. It is not a replacement for a traditional database. Remember, it's based on a document-storage concept. So, you must use it as a document storage with enhanced capabilities. Consider these different scenarios when thinking about how to store and partition your own content and choose the one that best matches your needs:

- **Big documents with almost flat structure** This is similar to object-oriented storage. It involves information that might have some nested elements (but not too many), but its definitions do not vary over time. A perfect example is a music library:

  - Each document represents a unit, like a CD, vinyl, or digital folder.

  - Each document has some attributes, like name, ID, year of publication, music style, and so on.

- Some sub-documents contain lists of performers in case there is more than one.
- There is a list of music themes.

- **Hierarchical structure**   If some of your data does not vary much over time, but some of it changes on a day-by-day basis, this would be the preferred schema. By using a parent key to identify the parent and another to identify the offspring, and having common data in both structures, you can simulate a relationship and store the information in small pieces. The COVID-19 sample data used for this book has this structure.

- **Mixed structure**   In this case, items under one partition key have pointers to child elements in another partition and vice versa. This can be applied when you have multiple elements with multiple dependents but the entire set of information is almost immutable. Consider as an example the same music library, but for a radio station, where they need to keep track not only of the media, but also the performers and their featured works. Each media item (CD, vinyl, and so on) contains an identifier for the performer. At the same time, each performer has an array of identifiers of the media items.

Implement a consistency model in Cosmos DB

Replicating data in several data centers and perhaps across several geographical regions requires you to manage the consistency of the data across the distribution. However, opting for the highest level of consistency slows the replication process down. So, you must consider whether increasing consistency will cause your applications to slow down too much during the replication process. It's important to strike the right balance.

Cosmos DB supports five different levels of consistency, explained in Table 2-4.

TABLE 2-4   Cosmos DB consistency levels

Consistency Level	Description
Strong	This level ensures that all the reads obtain the latest information committed.
Bounded staleness	The data has defined a staleness limit based on two parameters: quantity of versions and elapsed time. Because both parameters can be defined, the data will be consistent when either limit is reached. When data is read from the same region where it was updated, the engine ensures strong data consistency. Consistency varies depending on the type of account defined.
Session	A token identifies the writer and ensures that it and any others sharing the same token will get consistent prefix reads. This configuration also guarantees that any connection reading from the same region will get the most updated version, even when it is not the same connection, since they share the same token. For more on this, see the upcoming sidebar titled "The Session Consistency Level."
Consistent prefix	Only ordered updates are guaranteed. This means a reader will obtain information updated out of order. However, this does not mean the reader will obtain all the performed writes. It could get part of the set of writes, but always following the write order. For example, if a process writes three pieces of information, 1001, 1005, and 1007, the reader could retrieve only 1001 or 1001 and 1005 but never 1001 and 1007.
Eventual	This matches the generic consistency level. This means that eventually, the data will be synchronized across all the storage instances. This is not intended for transactional processes. It is for informational processes only, like those used on social networks to store messages, posts, replies, and so on, that do not need coherent grouping.

You can change the default consistency level for a database by using the Azure Portal (see the upcoming practice exercise) or by using various programmatic or scripting methods like PowerShell, Azure CLI, and .NET libraries.

Having a default consistency level does not mean you are tied to it. Programmatically, each call to the database can change the level, but only to a lower level. For example, when you create an instance of the DocumentClient class in .NET, which manages a Cosmos DB database, you can define the consistency for calls from that instance, as in the following code:

```
documentClient = new DocumentClient(
  endpointURI,
  authKey,
  connectionPolicy,
  ConsistencyLevel.Session);
```

Alternatively, you can change the consistency level for a single call like this:

```
RequestOptions reqOptions = new RequestOptions
        { ConsistencyLevel = ConsistencyLevel.Eventual };
var response = await documentClient.CreateDocumentAsync(
  collectionUri,
  documentToAdd,
  reqOptions);
```

The session consistency level

The session consistency level, which is the default for any new Cosmos DB account, uses a specific identifier for calls from a client application: the session token. You aren't required to manage the session token. The usual SDKs, like .NET and Java, manage it internally using the last one issued for each call. Each response to a call contains a token, which is used in the next call from the SDK. This means there is no direct relationship between multiple calls because the token changes with every response. However, if you want to manage a set of operations within the same session, you can manage the session token by yourself. To do so, you must obtain the session token from a response and keep using it during the subsequent requests. The following sample code shows you the steps to obtain and use the session token:

```
ItemResponse<Country> response =
    await container.CreateItemAsync<Country>(content);
string sessionToken = response.Headers.Session;
ItemRequestOptions options = new ItemRequestOptions();
options.SessionToken = sessionToken;
ItemResponse<Country> itemResponse =
    await container.ReadItemAsync<Country>(
        content.id,
        new PartitionKey(content.Code),
        options);
```

PRACTICE   Configure the default consistency level for a Cosmos DB

This practice exercise guides you through the steps to change the consistency model using Azure Portal.

1. Open *https://portal.azure.com* and use your credentials to log in.
2. Select your subscription, resource group, and Cosmos DB account.
3. In the left pane, under **Settings**, click **Default Consistency**.
4. Choose the desired consistency — **Strong**, **Bounded Staleness**, **Session**, **Consistent Prefix**, or **Eventual** — from the options along the top of the pane on the right.

   The map in the right pane changes depending on which option you choose to show the geographical distribution for your account configuration.
5. Move your mouse pointer over a region on the map.

   A message appears showing the replication delay for that region. (See Figure 2-4.)

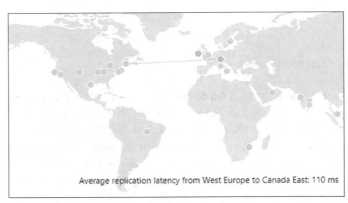

Average replication latency from West Europe to Canada East: 110 ms

FIGURE 2-4   Average replication between zones

Provision a non-relational data store

You've learned how to create a Cosmos DB account, database, and container using the Azure Portal. You've also learned how to create a storage account using the same method. But what about automating these processes?

You can do this using Azure Resource Manager (ARM) templates. An ARM template is a text file in Java Script Object Notation (JSON) format that describes the elements you want to create and the steps to perform the deployment. You can access ARM templates from the Azure Portal, by using scripting like PowerShell or Azure CLI, and even from .NET applications. You can also export ARM templates you create for later use.

PRACTICE   Download an ARM template from the Azure Portal

In this practice exercise, you will download an ARM template from the Azure Portal for a storage account.

1. Open *https://portal.azure.com* and use your credentials to log in.
2. Navigate to the storage account you created earlier in this chapter.

> **TIP**   You probably have a link to the storage account on the home dashboard. If not, just navigate to your resource group. You will find the account there.

3. In the left pane, near the bottom of the **Settings** group, click **Export Template**.

   The screen displays a directory tree for the ARM template associated with the current Azure storage account. To the right of the directory tree is a text editor that contains the ARM template's entire contents. This includes the following:

   - **Parameters**   This has one descendant node, which has the same name as the storage account.
   - **Variables**   This node is empty.
   - **Resources**   This node contains five descendant nodes: Account (Microsoft.Storage/storageAccounts), Blobs (Microsoft.Storage/storageAccounts/blobServices), Files (Microsoft.Storage/storageAccounts/fileServices), Queues (Microsoft.Storage/storageAccounts/queueServices), and Tables (Microsoft.Storage/storageAccounts/tableServices).

4. Click the **Download** button above the directory tree.

   The template is downloaded and saved in compressed ZIP format.

5. Extract the contents of the ZIP file, which consist of two files: parameters.json and template.json.

> **NOTE**   The parameters.json file enables you to use different parameter values in different executions to avoid having to edit the template file each time.

Analyze the ARM template

The ARM template consists of typical JSON content. It starts by establishing what schema to use, as follows:

```
"$schema": "https://schema.management.azure.com/schemas/2019-04-01/
deploymentTemplate.json#",
    "contentVersion": "1.0.0.0",
```

After that, the JSON file contains a `parameters` entry, where parameters used by the template are declared:

```
"parameters": {
    "storageAccounts_ads_saccount_name": {
        "defaultValue": "adssaccount",
        "type": "String"

    }

},
```

The `parameters` entry is where you change the default values to deploy other accounts of the same type. When a parameters file is used, the values in it replace the default values declared here.

The next node is for variables. In this case, it is empty.

After that, the resources to be created are defined, starting with the storage account itself:

```
{
    "type": "Microsoft.Storage/storageAccounts",
    "apiVersion": "2019-06-01",
    "name": "[parameters('storageAccounts_adssaccount_name')]",
    "location": "northeurope",
    "sku": {
        "name": "Standard_LRS",
        "tier": "Standard"
    },
    "kind": "StorageV2",
    "properties": {
        "minimumTlsVersion": "TLS1_2",
        "allowBlobPublicAccess": false,
        "isHnsEnabled": true,
        "networkAcls": {
            "bypass": "AzureServices",
            "virtualNetworkRules": [],
            "ipRules": [],
            "defaultAction": "Allow"
        },
        "supportsHttpsTrafficOnly": true,
        "encryption": {
            "services": {
                "file": {
                    "keyType": "Account",
                    "enabled": true
                },
                "blob": {
                    "keyType": "Account",
                    "enabled": true
```

```
                }
            },
            "keySource": "Microsoft.Storage"
        },
        "accessTier": "Hot"
    }
},
```

Table 2-5 lists the most important parts of the ARM template.

TABLE 2-5   ARM template entries

Name	Mandatory	Description
Type	✓	Defines the resource type. This entry must include the resource provider (called *namespace*), a slash, and the name of the resource to create, like so: `Microsoft.Storage/storageAccounts`.
apiVersion		Contains the version for the REST API to use. Usually, the version is a date in ISO 8601 format (yyyy-mm-dd). However, it can be extended to include some other data, like in the last REST API version for databases (2019-06-01-preview). If no value is provided, the value for the `apiVersion` property for the entire template will be used.
name	✓	Contains the name of the resource to be created. This name must be defined as a URI, following the RFC3986 specification, because it could be exposed publicly and must match the rules. The name must also be unique. During resource deployment, a process will validate the uniqueness of the name inside Azure and, if it will be exposed publicly, outside of Azure.
location		The geographical location. Several resources, but not all, require this. This value should match the location of other resources in the same subscription and resource group or other resources that will communicate frequently with this new one. For example, you could define the location as parameter in your ARM template and set its default value to match the one for the resource group like so: `"defaultValue": "[resourceGroup().location]"`
dependsOn		Used to declare other resources cited in the ARM template as a comma-separated list of resource names or identifiers. These other resources, which are external to the template, must be created before the template is deployed. Otherwise, the template will not be able to check for the resources. The deployment process analyzes the `dependsOn` attribute for all the resources defined. It then sorts them to first create those without dependencies (in parallel) and then follow the dependency chains to complete the deployment without failures.

sku		Defines the specific version or style for the resource to create. For example, in a storage account, the `name` parameter defines the replication type, and the `tier` parameter defines the performance level, like so:
		
		 `"sku": {`
		
		 `"name": "Standard_LRS",`
		 `"tier": "Standard"`
		 `}`

kind		Defines other resource specifics. Some resources require this value for their definition. In the case of a storage account, `kind` defines the type of storage — in this example, `StorageV2`.

Name	Mandatory	Description

properties		Defines specific values for the resource. Each resource will have its own schema. For example, a storage account, will have these properties:
		
		``json
		"properties": {
		    "minimumTlsVersion": "TLS1_2",
		    "allowBlobPublicAccess": false,
		    "isHnsEnabled": true,
		    "networkAcls": {
		        "bypass": "AzureServices",
		        "virtualNetworkRules": [],
		        "ipRules": [],
		        "defaultAction": "Allow"
		    },
		    "supportsHttpsTrafficOnly": true,
		    "encryption": {
		        "services": {
		"file": {
		    "keyType": "Account",
		    "enabled": true
		},
		"blob": {
		    "keyType": "Account",
		    "enabled": true
		}
		        },
		        "keySource": "Microsoft.Storage"
		    },
		    "accessTier": "Hot"
		}
		``
		
		Configurations for networking, secure access, encryption, and the access tier are also defined here.

PRACTICE   Change the ARM template for storage

You can edit the template to better fit your requirements. For example, you can add parameters for location, SKU, or other properties. To the change the ARM template for storage follow these steps:

1. With the text editor of your preference, open the **template.json** file you downloaded previously.

> *NOTE* For this practice exercise, you can use any text editor you like. However, if your work requires frequent manipulation of ARM templates, we recommend you use Visual Studio Code, available for installation here: *https://code.visualstudio.com/Download*. Versions for Windows, Linux, and MacOS are available. For more information about this tool, see the sidebar at the end of this practice exercise.

2. Add a parameters definition for the SKU called `skuname`. To do so, replace the existing parameters section with the following code. (Notice that you can define allowed values here as well.)

```
"parameters": {
    "accountName": {
        "defaultValue": "adssaccount",
        "type": "String"
    },
    "skuname": {
        "type": "String",
        "defaultValue": "Standard_LRS",
        "allowedValues": [
            "Standard_LRS",
            "Standard_GRS",
            "Standard_RAGRS",
            "Standard_ZRS",
            "Premium_LRS",
            "Premium_ZRS",
            "Standard_GZRS",
            "Standard_RAGZRS"
        ]
    }
},
```

3. Search for all occurrences of the old name of the storage account and replace them with the new name. In this example, that means replacing instances of `storageAccounts_ adssaccount_name` with `accountName`.

4. Repeat step 3 for the SKU. In this example, this likely means replacing instances of `Standard_LRS` with `[parameters('skuname')]`.

5. Change the resource location to match the resource group's by using an environment variable called `resourceGroup().location`. To start, find the first instance of the `location` argument — probably in the storage account definition. Then, using your text editor to search and replace, replace each instance of the argument with the environment variable. Be sure the environment variable is delimited by double quotes, as in the following:

```
"location": "[resourceGroup().location]",
```

6. Save the template file.

7. Open the **parameters.json** file.

8. Change the `parameters` values for the next deployment to something like the following to match the new definitions:

```
"parameters": {
    "accountName": {
        "value": "adssa2"
```

```
    },
    "skuname": {
        "value": "Standard_LRS"
    }
}
```

Using Visual Studio Code

If your work requires frequent manipulation of ARM templates, we recommend that you install Visual Studio Code. After you do, open it and press **Ctrl+Shift+X** to look for extensions. Then search for and install the **Azure Resource Manager (ARM) Tools** extension. This enhances your editing experience by adding Intelli-Sense to the editor for ARM templates.

Visual Studio Code also makes it easier to access the entire contents of the down-loaded ARM template. Just open the **File** menu, choose **Open Folder**. Then open the downloaded ZIP file and open the resulting folder. Figure 2-5 shows how the contents of the folder look in Visual Studio Code, as well as the template.json file.

FIGURE 2-5 Visual Studio Code with a folder open and its contents showing

Automate resource deployment

After you create and save a customized template, you can use it and the parameters file to deploy new resource instances at any time using various tools available for Azure resource management. These include the following:

- PowerShell
- The command line
- The .NET Framework, including .NET Core

> **NOTE** In addition to creating custom templates, you can find hundreds of pre-defined templates called Azure Quickstart templates here: *https://azure.microsoft.com/en-us/resources/templates/*.

USE POWERSHELL TO DEPLOY A NEW RESOURCE

To perform a deployment using PowerShell, you must first import the Az module to your environment. If necessary, you must also uninstall a legacy library called AzureRM. To do this, use the following script. (This script is also available in the companion content for this book, in a file called Install az Module.ps1.)

```
$version=$PSVersionTable.PSVersion.ToString()
if($version-lt'5.1.0'  ){
    Write-Warning -Message ('az Module requires Powershell version 5.1 or higher.'+
    ' Your version is $version. '+
    'Go to https://docs.microsoft.com/en-us/powershell/scripting/install/"+
    "installing-powershell to install the latest version')
 }
else {
    if ($PSVersionTable.PSEdition -eq 'Desktop' -and (Get-Module -Name AzureRM
-ListAvailable)) {
        Write-Warning -Message ('Az module not installed. Having both the AzureRM and ' +
           'Az modules installed at the same time is not supported.'+
           'Follow the instructions at '+
           'https://docs.microsoft.com/en-us/powershell/azure/"+
           "uninstall-az-ps?view=azps-4.3.0#uninstall-azure-powershell-msi'+
            'to remove it before install az Module' )
    }
    else {
        Install-Module -Name Az -AllowClobber -Scope CurrentUser
    }
}
```

With the Az module installed, you can use the PowerShell script shown in Listing 2-1 to deploy a new storage account. Notice that this script uses parameters to set the resource group name and path for the template and parameters file. (This code is also available in the Deploy Storage Account.ps1 file with the companion content.)

LISTING 2-1   Deploy storage account

```
[CmdLetBinding()]
param
(
    [Parameter(Mandatory=$true,Position=0,HelpMessage="You must set the Template file
path" )]
        [string]$templateFile,
    [Parameter(Mandatory=$true,Position=1,HelpMessage="You must set the Parameters file
path" )]
        [string]$parameterFile,
    [Parameter(Mandatory=$true,Position=2,HelpMessage="The Resource Group name must be
defined" )]
        [string]$resourceGroupName
)
Clear-Host
$badParameters=$false
if  (-not (Test-Path $templateFile -PathType leaf))
{
   Write-Host "ERROR: The template file $templateFile do not exist" -ForegroundColor Red
   $badParameters=$true
}
if  (-not (Test-Path $parameterFile -PathType leaf))
{
   Write-Host "ERROR: The parameters file $parameterFile do not exist" -ForegroundColor Red
   $badParameters=$true
}
<#
    NOTE: remove the #» to execute the commands
    to check and /or install the modules
#>
#                   Check you have the Azure Snap-in installed in your environment
      #»Get-InstalledModule -Name Az
#                   If there is no module for az, install it
      #Install-Module -Name Az -AllowClobber -Force
#Load the Azure Snap-in
Import-Module -Name Az
#Connect to your Azure Account
Connect-AzAccount
#Check if Resource Group exists
$rg=Get-AzResourceGroup -Name $resourceGroupName
if($rg -eq $null)
{
   Write-Host "ERROR: The Resource Group $resourceGroupName "+
      "do not exist in the current subscription" -ForegroundColor Red
   $badParameters=$true
}
```

```
if($badParameters)
{
    return
}

#Deploy the Template
New-AzResourceGroupDeployment `
    -ResourceGroupName $resourceGroupName `
    -TemplateFile $templateFile `
    -TemplateParameterFile $parameterFile
```

USE THE COMMAND LINE TO DEPLOY A NEW RESOURCE

With the Azure CLI, you can deploy the same template and parameters file by issuing the following commands using a command window or Windows Terminal, with bash in Linux, and from the Azure Cloud Shell:

```
az login
az deployment group create \
    --resource-group resourceGroupName \
    --template-file templateFileName.json \
    --parameters parametersFileName.json
```

> **NOTE**   The Azure Cloud Shell is a CLI environment available from the Azure Portal. You can use it to preserve and execute scripts, templates, and so on. When using the Azure Cloud Shell, you can switch between using bash and PowerShell. The Azure Cloud Shell also contains an editor to facilitate the creation of your own scripts. For more details, see *https://docs.microsoft.com/en-us/azure/cloud-shell/overview*.

USE THE .NET FRAMEWORK TO DEPLOY A NEW RESOURCE

The Azure Management API contains a complete set of .NET libraries to deploy your own code using various operating systems. In the Azure.ResourceManager.Resources.Models library, a `Deployment` class defines a new deployment by configuring the sources in its `Properties` property as `DeploymentProperties` type.

> **NOTE**   For resource deployments, you need the Azure.ResourceManager.Resources package, which at the time of this writing is still in pre-release. In addition, to enable authentication, you must use the Azure.Identity package.

There are some tricks to using the parameters file with .NET libraries. For example, you must extract from the parameters file only the segment that describes the parameters. You must then convert this segment to a string to create a `DeploymentOptions` class instance. You can see how to do this in the following code, from the Deploy ARM Template Visual Studio sample in the companion content.

```
// Generate a random deployment name
string deploymentName = "dep" + DateTime.Now.ToFileTime().ToString();
// Read the content of the template file into a string
string templateJson = File.ReadAllText(templateFile);
// Get the Parameters from the JSON file
ParametersObject parametersobj =

System.Text.Json.JsonSerializer.Deserialize<ParametersObject>
(File.ReadAllText(parametersFile));
// Convert the name value pairs of the parameters into a JSON string
string parametersJson =
System.Text.Json.JsonSerializer.Serialize(parametersobj.parameters);
var deployments = resourceClient.Deployments;
var parameters = new Deployment
    (
        new DeploymentProperties(DeploymentMode.Incremental)
        {
            Template = templateJson,
            Parameters = parametersJson
        }
    );
var rawResult = await deployments.StartCreateOrUpdateAsync(
    resourceGroupName, deploymentName, parameters);
await rawResult.WaitForCompletionAsync();
```

Note that this code uses a `ParameterObject` class, defined as follows:

```
/// <summary>
/// This class represents the content of the parameters.json file, according to the schema at
/// https://schema.management.azure.com/schemas/2015-01-01/deploymentParameters.json
/// </summary>
public class ParametersObject
{
  public string schema { get; set; }
  public string contentVersion { get; set; }
  /// <summary>
  /// The parameters property is defined of object type,
  /// to receive any quantity and combination of parameters.
  /// When serialized, the result string is the exact schema definition
  /// the <see cref="DeploymentProperties.Parameters"/> expect.
  /// </summary>
  public object parameters { get; set; }
}
```

Add content to your storage

This section contains several exercises that show you how to use different tools to upload data to storage. First, though, we want to talk briefly about the sample data used here.

As you know, many people have been affected by the COVID-19 virus. At first, scientists did not know much about the disease. Universities, research centers, and healthcare facilities quickly collected information for analysis purposes and to try to predict how the virus would behave. The exercises in this section use some of this public data — modified slightly to facilitate their use. The data set contains general COVID 19–related information for countries around the world and day-by-day case reports.

The general information for each country is in a JSON file called `covid-Data-Countries.json`. The following code is one entry in that file. (In keeping with the previously defined container, the partition key is "Code".)

```
{
 "Code": "Country",
 "continent": "Europe",
 "location": "Albania",
 "population": 2877800,
 "population_density": 104.871002,
 "median_age": 38,
 "aged_65_older": 13.1879997,
 "aged_70_older": 8.64299965,
 "gdp_per_capita": 11803.4307,
 "cardiovasc_death_rate": 304.195007,
 "diabetes_prevalence": 10.0799999,
 "handwashing_facilities": 0,
 "hospital_beds_per_thousand": 2.8900001,
 "life_expectancy": 78.5699997,
 "data": null,
 "CountryCode": "ALB"
}
```

Notice that the Code attribute contains the value "Country". The reports for each country are in different files, one per country, with the following structure:

```
{
 "date": "2020-03-13",
 "total_cases": 2,
 "new_cases": 2,
 "total_deaths": 0,
 "new_deaths": 0,
 "total_cases_per_million": 18.7329998,
 "new_cases_per_million": 18.7329998,
 "total_deaths_per_million": 0,
 "new_deaths_per_million": 0,
 "stringency_index": 0,
```

```
 "new_cases_smoothed": 0,
 "new_deaths_smoothed": 0,
 "new_cases_smoothed_per_million": 0,
 "new_deaths_smoothed_per_million": 0,
 "Code": "Data",
 "Country": " ALB "
}
```

In these entries, the Code attribute contains the value "Data". When the container was created, you defined the Code attribute as the partition key. So, the data, which consists of detailed COVID information for various countries, will be stored in two different partitions: one for countries and the other for data.

> **NOTE** This data is available in the Data folder for Chapter 2 in this book's companion content.

EXERCISE 2-1   Feed a Cosmos DB container using JSON source files

In this exercise, you will upload JSON files to a Cosmos DB container. Follow these steps:

1. Open *https://portal.azure.com* and use your credentials to log in.
2. Navigate to your Cosmos DB account.
3. Click **Data Explorer** in the pane on the left.
4. Expand the directory tree that appears to locate your container.
5. Click the **Items** node to select it.
6. Click the **Upload Item** button above the panel to the right.
7. In the **Upload Items** pane, locate and select the **covid-Data-Countries.json** file. (See Figure 2-6.)

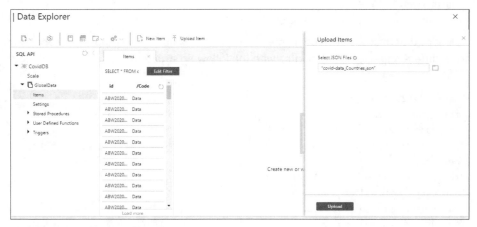

FIGURE 2-6   Upload data to Cosmos DB

When the upload process is complete, you will see one entry per country in the Items list.

8. Click a country in the **Items** list to see the corresponding data in JSON format.

   The following attributes are automatically added by Cosmos DB as internal properties, with different values for each item:

```
"id": "56411784-8c51-4ef0-a45b-8b85b869ed0c",
    "_rid": "aCBpAJKWogUHhR4AAAAAA==",
    "_self": "dbs/aCBpAA==/colls/aCBpAJKWogU=/docs/aCBpAJKWogUHhR4AAAAAA==/",
    "_etag": "\"0300939b-0000-0d00-0000-5f5be3810000\"",
    "_attachments": "attachments/",
    "_ts": 1599857537
```

9. Repeat the upload process but this time select one or several files in the **covid-data_Data.<country_Code>.json** list.

   > **NOTE** You cannot upload more than 2 MB by set, so limit your selection to 10 or so files to avoid stopping the process.

   The visible data corresponds to entries with a Code attribute value of "Country". So, to look at data items, you must use a filter.

10. To search for data items, click the **Edit Filter** button.

11. Add the following code to the filter and click the **Apply Filter** button to filter the data entries.

```
WHERE c.Code = "Data"
```

EXERCISE 2-2   Create a container in Azure storage using Portal Storage Explorer

There are different tools to upload data into an Azure storage account. All of them will accomplish the task. The difference is in how you use them to insert the data. In this exercise, you will use Portal Storage Explorer to create a container and upload data. Follow these steps:

1. Open *https://portal.azure.com* and use your credentials to log in.

2. Navigate to your storage account.

3. Click **Storage Explorer** in the pane on the left.

   A directory tree appears with nodes for all the different kinds of storage you can manage in the account and the elements created within them. To the right is a pane that lists the content within the selected node. (See Figure 2-7.)

FIGURE 2-7   Storage Explorer

4. Right-click the **Containers** node in the directory tree and select **Create File System** from the menu that appears.

5. Type a name for the container in lowercase letters — in this example, **globaldata** — and click **OK.**

6. Click the new **container** in the directory tree to select it.

7. Click the **New Folder** button above the right pane and add a **Demographic** folder.

EXERCISE 2-3   Add content to Azure storage using Azure Storage Explorer

Although there is an Upload button in the Portal Storage Explorer screen, it has recently been disabled, meaning you can no longer use it to upload content. Instead, you must use another tool: Azure Storage Explorer.

Azure Storage Explorer is a multiplatform application. You can install it on your Windows, MacOS, or Linux system from here: *https://azure.microsoft.com/en-us/features/storage-explorer/.* After you install it, enter your usual Azure credentials to retrieve the storage accounts from all your active subscriptions. (These can be enabled or disabled from the directory tree at any time.) Then follow these steps to add content to a storage account.

1. Select the storage account and container to which you want to add content.

   You will see the same interface you saw in Portal Storage Explorer.

2. Expand the **Containers** node in the directory tree and select the previously created container — in this example, **globaldata.**

3. Double-click the **Demographic** folder you created in the previous exercise.

4. Click the **Upload** button and select **Upload Files**.

5. In the **Upload Files** dialog box, locate and select the **covid-data_Countries.json** file in the **\Data** folder in the Chapter 2 companion content.

6. Confirm that the destination folder is **/Demographic/** and click the **Upload** button. (See Figure 2-8.)

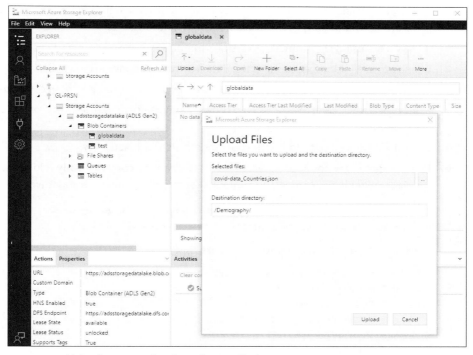

FIGURE 2-8 Upload content using Azure Storage Explorer

7. To view the contents of an uploaded file, double-click the file in the pane on the right. Alternatively, select the file and click the **Open** button.

   If you already have an application to view JSON files installed on your system, the file will be opened by that application. If not, your operating system will ask you which application you want to use.

EXERCISE 2-4   Add content to Azure storage using AzCopy

AzCopy is a command-line utility for performing storage operations. The latest and recommended version at this time is V10, which is available here: *https://docs.microsoft.com/en-us/azure/storage/common/storage-use-azcopy-v10*.

> **NOTE**   The preceding link also contains information about how to use the utility as well as security requirements — user roles, authentication, and so on. For example, to upload files, the user must be assigned the **Storage Blob Data Contributor** or the **Owner role**.

After you install AzCopy, follow these steps to create a new folder and upload a set of files:

1. Open a command window.

   - If you are using Windows, open a command or Windows Terminal window.

- If you are using macOS, locate the Terminal application in the Applications or Utilities folder or by using any of the available search methods.
- If you are using Linux, use your version's method to open a command prompt.

2. Change to the folder that contains the data from the companion content.

3. To authenticate the user, use the `<Path_To_azCopy_Folder>\azcopy login` command.

4. Follow the instructions to open a Web browser and navigate to *https://microsoft.com/devicelogin*.

5. Enter the displayed code to match your credentials in the AzCopy window.

6. In the command window, use the following command to copy the **Daily** folder inside the **Data** folder in the companion content. (Replace the values with your own.)

```
<Path_To_azCopy_Folder>\azcopy copy 'Daily' 'https://<storage-account-name>.blob.core.windows.net/<container-name>/<blob-name>' --recursive
```

The results summary will look somewhat like this:

```
Job cd4bc9de-3f6f-1840-59a0-10ca3a883b27 summary
Elapsed Time (Minutes): 0.1667
Number of File Transfers: 211
Number of Folder Property Transfers: 0
Total Number of Transfers: 211
Number of Transfers Completed: 211
Number of Transfers Failed: 0
Number of Transfers Skipped: 0
TotalBytesTransferred: 18155562
Final Job Status: Completed
```

> **NOTE** The `--recursive` modifier is a tool to upload content as a hierarchical structure for ADLS.

7. Use Azure Storage Explorer or Portal Storage Explorer to review the Blob storage and confirm that it contains a Daily folder with 211 files.

USE APIS FOR MORE COMPLEX UPLOADING TASKS

If you need more control or to perform more complex uploading tasks, you can use other APIs. For example, you can upload files using a PowerShell script. (As mentioned, you need the Azure PowerShell module to perform Azure actions from PowerShell. You can use the Install az Module.ps1 script in the companion content to do this.)

Listing 2-2 provides a sample PowerShell script to perform the upload you already did, including the folder creation.

LISTING 2-2 PowerShell script to upload files to blobs

```
<#
    NOTE: remove the #» to execute the commands
    to check and /or install the modules
#>
#                       Check you have the Azure Snap-in installed in your environment
        #»Get-InstalledModule -Name Az
#                       If there is no module for az, install it
        #Install-Module -Name Az -AllowClobber -Force
#Load the Azure Snap-in
Import-Module -Name Az
#Connect to your Azure Account
Connect-AzAccount
#Define the Variables
$resourceGroupName= "<your_Resource_Group>"
$accountName = "<your_Storage_Account_Name>"
$containerName="<your_Container_Name>"
$DataFolder="<path_to_your_local_Data_folder_From_Companion_Content>"
$storageAccount=Get-AzStorageAccount -ResourceGroupName $resourceGroupName `

    -Name $accountName
$storageContext=$storageAccount.Context
$countriesFileName="$DataFolder\covid-data_Countries.json"
Clear-Host
Write-Host "Uploading [Countries] data"
#Upload the Countries file to the Demography folder
$dummy=Set-AzStorageBlobContent `
    -Container $containerName `
    -File $countriesFileName `
    -Blob "Demography\Countries" `
    -Context $storageContext
$Files=get-item "$DataFolder\Daily\*.*"
#Upload all the daily data files to the Daily folder
foreach($oneFile in $Files){
    #Remove unnecessary text from the name, leaving only the country code as blob name
    $name=$oneFile.Name.Replace("covid-data_Data.","").Replace(".json","")
    Write-host "Uploading data for [$name]"
    $dummy=Set-AzStorageBlobContent `
        -Container $containerName `
        -File $oneFile `
        -Blob "Daily\$name" `
        -Context $storageContext
}
```

Notice that this code cleans the file names to make them more readable. Of course, you can use the original names or any other naming convention you prefer, which is more difficult to do using the other methods.

USE MICROSOFT.AZURE.STORAGE.BLOB TO INTEGRATE CONTENT UPLOAD PROCEDURES INTO AN APP

If you want to integrate content upload procedures into an app, you can do so using .NET, as shown in Listing 2-3. This code is a class sample that is useful for performing uploads to blobs programmatically.

> **NOTE** The code requires you to add `Microsoft.Azure.Storage.Blob` via NuGet packages.

LISTING 2-3 C# class to upload content to Azure Blob storage

```csharp
using Microsoft.Azure.Storage.Blob;

using System;
using System.IO;
using System.Threading.Tasks;

namespace ADS.Azure
{
  public static class Blob
  {
    private static CloudBlobClient blobClient;
    static string AZBlobContainer;
    static Blob()
    {
    // This could be loaded from your configuration
ConfigurationManager.AppSettings["AZStoreaccountName"];
      string AzStoreAccountName = "<>";
                // This could be loaded from your configuration
ConfigurationManager.AppSettings["AZBlobContainer"];
      AZBlobContainer = "<>";

        Uri uri = new Uri($"https://{AzStoreAccountName}.blob.core.windows.net");
        // This could be loaded from your configuration
ConfigurationManager.AppSettings["AZStorageKey"]
      string AZStorageKey = "";
Microsoft.Azure.Storage.Auth.StorageCredentials credentials =
          new Microsoft.Azure.Storage.Auth.StorageCredentials(
AzStoreAccountName, AZStorageKey);
      blobClient = new Microsoft.Azure.Storage.Blob.CloudBlobClient(uri, credentials);
    }
    /// <summary>
    /// Uploads a content to a blob.
    /// </summary>
    /// <param name="blobName">Name of the blob.</param>
    /// <param name="content">The content to upload as string.</param>
    /// <param name="folderName">Name of the destination folderfolder.</param>
```

```csharp
        static public async Task UploadFileAsync(
            string blobName,
            string content,
            string folderName = "/")
        {
            try
            {
                CloudBlobContainer folder = blobClient.GetContainerReference(AZBlobContainer);
                CloudBlobDirectory directory = folder.GetDirectoryReference(folderName);

                CloudBlockBlob file = directory.GetBlockBlobReference(blobName);
                await file.UploadFromStreamAsync(GenerateStreamFromString(content));

            }
            catch (Exception)
            {
                throw;
            }

        }
        /// <summary>
        /// Uploads a content to a blob.
        /// </summary>
        /// <param name="blobName">Name of the blob.</param>
        /// <param name="content">The content to upload as string.</param>
        /// <param name="folderName">Name of the destination folderfolder.</param>
        static public void UploadFile(
          string blobName,
          string content,
          string folderName = "/")
        {
          (Task.Run(() =>
              UploadFileAsync(blobName, content, folderName))
          ).Wait();
        }
        #region "Helpers"
        static Stream GenerateStreamFromString(string s)
        {
            var stream = new MemoryStream();
            var writer = new StreamWriter(stream);
            writer.Write(s);
            writer.Flush();
            stream.Position = 0;
            return stream;
        }
        #endregion "Helpers"
    }
}
```

Provision an Azure Synapse Analytics workspace

Azure Synapse Analytics is a business intelligence (BI) implementation for data analysis. It was originally based on Azure SQL Data Warehouse but changed to better integrate with multiple data-analysis technologies and to use massive storage tools like ADLS. For example, data can be prepared as Parquet tables, which can be processed by several engines, including Spark and Synapse SQL.

Parquet

Parquet is a storage format developed in 2013 by Twitter and Cloudera for Hadoop. It is based on the column storage concept to enhance compression and query performance. Information is not stored in Parquet on a row-by-row basis, where each row could have different data types. Instead, it groups all the columns of the various entries, which are better suited for analytics. There are two main reasons for this:

- Each column store contains just one data type. This allows better compression algorithms.
- Values are stored in order, so it is easier to perform lookups and group them in ranges.

The Parquet format is defined by messages, which must have three attributes:

- **Repetition** This can be one of the following values:
  - **Required** There must be one entry for this column.
  - **Optional** There can be 0 or 1 columns.
  - **Repeated** There can be 0 to many values.
- **Name** This must be unique in the same level.
- **Type** This defines the data type to store. Table 2-6 lists the data types used in Parquet.

TABLE 2-6 Parquet data types

Type	Description
DATE	The date, not including the time of day. It uses int32 annotation. It stores the number of days from the Unix epoch (1 January, 1970).
DECIMAL	An arbitrary-precision signed decimal number in the form *unscaledValue * 10^(-scale)*.
INT_8	8 bits
INT_16	16 bits
INT_32	32 bits
INT_64	64 bits

Type	Description
INTERVAL	An interval of time. It annotates a fixed-length byte array of length 12. Months, days, and milliseconds are presented in unsigned little-endian encoding.
TIME_MILLIS	The logical time, not including the date. It annotates in an INT32 data type. It refers to the number of milliseconds since midnight.
TIMESTAMP_MILLIS	The logical date and time. It annotates in an INT64 data type that stores the number of milliseconds from the Unix epoch, 00:00:00.000 on 1 January 1970, UTC.
UINT_8	8 bits unsigned
UINT_16	16 bits unsigned
UINT_32	32 bits unsigned
UINT_64	64 bits unsigned
UTF8 (strings)	Annotates the binary primitive type. The byte array is interpreted as a UTF-8 encoded character string.
GROUP	Contains groups of elements defined by another message.

NOTE The GROUP data type enables Parquet to store hierarchical information in multiple levels.

NEED MORE REVIEW? For a complete explanation of the Parquet format, see *https:// blog.twitter.com/engineering/en_us/a/2013/dremel-made-simple-with-parquet.html*. To find the latest on Parquet updates, see *http://parquet.apache.org/*.

PRACTICE Implement an Azure Synapse Analytics workspace

In this practice exercise, you'll implement an Azure Synapse Analytics workspace, which will be useful for other practice exercises in this book.

1. Open *https://portal.azure.com* and use your credentials to log in.

2. In the search box, type **Azure Synapse Analytics**, and select the corresponding entry that appears in the drop-down list.

3. Click the **Create Synapse Workspace** button. Alternatively, click the **plus sign** in the toolbar above the list of existing workspaces.

4. In the **Basics** tab, configure the following settings:

   - **Subscription** Select your subscription.
   - **Resource Group** Select your resource group.
   - **Managed Resource Group** Type a name for a managed resource group that is specific to the workspace. This is created to store ancillary resources during data processing. You can specify a name to match your standards or leave it empty, in which case an auto-generated name will be used.

- **Workspace Name**   Type a name for the new workspace.
- **Region**   Choose a region. (The default region is the one associated with the resource group you selected.)

5. Under **Select Data Lake Storage Gen2**, use the **Account Name** settings to choose the ADLS Gen2 account you want to use or to create a new one.

> **NOTE**   The ADLS Gen2 account you choose must have hierarchical name space (HNS) enabled.

6. Use the **File System** settings to choose the file system (container) you want to use or to create a new one.

> **TIP**   Because the Azure Synapse Analytics workspace is not a repository for your data, but rather a storage for catalogs and metadata, it is a good idea to create a new file system to isolate it from your analytics data.

7. Click the **Assign Myself the Storage Blob Data Contributor Role...** check box to select it.
8. Click the **Security + Networking** tab and configure the following settings:
   - **Admin Username**   Type a user name for the administrator account. (It is a best practice to change the default one.)
   - **Password/Confirm Password**   Type the password you want to use for the account.

> **NOTE**   Do not use double encryption with a customer-managed key unless you already have keys in the subscription's Key Vault and you want to implement them.

   - **Allow Pipelines...to Access SQL Pools**   Leave this check box selected.
   - **Allow Connections from All IP Addresses**   Leave this check box selected. You will review how to limit this in Chapter 4, "Monitor and optimize data solutions."
   - **Enable Managed Virtual Network**   Select this check box if you need to implement connections via Azure Private Link.

9. Click the **Tags** tab and enter any custom tags you like.
10. Click the **Review + Create** tab. Then click the **Create** button.

   You will see the approximate cost based on your configuration settings and the region you selected.

After the workspace has been created, it will be listed in your workspace group as a Synapse workspace, with the following resources configured:

- Microsoft.Resources/deployments
- Microsoft.Synapse/workspaces/firewallrules
- Microsoft.Synapse/workspaces/managedIdentitySqlControlSettings
- Microsoft.Synapse/workspaces
- Microsoft.Synapse/workspaces
- Microsoft.Resources/deployments

PRACTICE   Add an Apache Spark pool

Creating a workspace automatically defines a serverless SQL pool for your work. However, you can create a dedicated SQL pool or a serverless Apache Spark pool if you prefer. This practice exercise guides you through the creation of an Apache Spark pool in your workspace, which you can use in future practice exercises.

1. Click the **Open Synapse Studio** link on the workspace **Overview** page.

   Synapse Studio opens, with the parameters for your subscription and workspace ID already set.

   > **NOTE** You use Synapse Studio to manage your workspace and to ingest and process data.

2. In Synapse Studio, click the **Manage** button in the left toolbar.
3. Under **Analytic Pools**, click **Apache Spark Pools**.
4. Click the **New** button.
5. In the **Basics** tab, configure the following settings:
   - **Apache Spark Pool Name**   Type a name for the new pool.
   - **Autoscale**   Leave this set to **Enabled**.
   - **Node Size**   For this practice exercise, select the lowest size value.
   - **Number of Nodes**   Leave this at the default value of **3**.
6. Click the **Additional Settings** tab and configure the following settings:
   - **Auto-Pause**   Leave this set to **Enabled**.
   - **Number of Minutes Idle**   Leave this set at the default value of **15**.
   - **Apache Spark**   Choose the appropriate version of Apache Spark. (At the time of this writing, there is just one option: **2.4**.)
7. Review the versions for the different languages and tools that are automatically enabled.

   > **NOTE** You can upload Apache Spark configuration files if desired.

8. Click the **Tags** tab and enter any custom tags you like.

9. Click the **Review + Create** tab. Then click the **Create** button.

10. Click the name of the new pool. You'll have the option of being directed to the Azure Portal to see the deployment status.

PRACTICE   Add a dedicated SQL pool

This practice exercise guides you through the creation of a dedicated SQL pool in your workspace, which you can use in future practice exercises.

1. In the workspace **Overview** page, click the **New SQL Pool** button.

2. In the **Basics** tab, configure the following settings:

   - **SQL Pool Name**   Type a name for the new dedicated SQL pool.
   - **Performance Level**   For this practice exercise, use the slider to set the performance level to the lowest value.

> **NOTE**   You can also see any SQL pools that have already been created on this page.

> **NOTE**   The page updates the cost information when you change the Performance Level setting.

3. Click the **Additional Settings** tab and configure the following settings:

   - **Use Existing Data**   Set this to **None**.
   - **Collation**   Leave this set at the default value.

4. Click the **Tags** tab and enter any custom tags you like.

5. Click the **Review + Create** tab. Then click the **Create** button.

Provide access to data to meet security requirements

It's critical that your data in the cloud is protected. Fortunately, there are a lot of security procedures managed internally by Microsoft teams to ensure this data is secure. However, securing your data is a joint responsibility. On the one hand, the Azure teams at Microsoft ensure all the needed infrastructure security is in place. On the other, it is your responsibility to configure your resources using the least-permissions principle.

> **NEED MORE REVIEW?**   For detailed information about how security is implemented, see *https://docs.microsoft.com/en-us/azure/security/fundamentals/infrastructure*.

As you can see in Figure 2-9, different kinds of users and resources can access Azure assets. (This is why Microsoft refers to them not as *users*, but as *identities*.) These resources could be other services working inside Azure platforms or external to them, including on-premises applications, services, and so on.

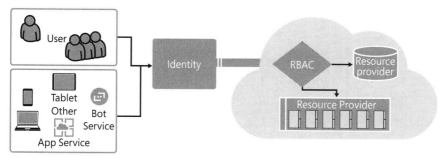

FIGURE 2-9   Azure security

Azure identifies resources using specific data. For example, a storage account could be accessed using a storage key. Azure must identify a resource before allowing it access. In a similar way, users, being actual people, must be identified as well.

There are various components to identification. These include the following:

- Azure Active Directory (AD)
- Role-based access control
- Other authorization methods

Azure Active Directory

Azure uses Azure Active Directory (AD) to manage all identities. This is similar to the Active Directory that appeared first in Windows 2000 Server, except for the use of organizational units, which are not part of it. By default, all Azure subscriptions have AD, where users and groups can be created and their permissions managed.

Office 365 and Azure AD

Office 365 subscriptions uses the same AD as Azure, which is managed by Azure. This enables users with an O365 subscription to access other resources in Azure subscriptions. Moreover, the management of the O365 memberships points to the Azure AD directly.

You access the Azure AD Administration Portal at *https://portal.azure.com/#blade/ Microsoft_AAD_IAM/ActiveDirectoryMenuBlade/Overview* (see Figure 2-10). (Not all options will be available for all users, and not all users will have the necessary permissions to make changes in this portal.) Here, you can assign several security levels to users to restrict or allow various actions and views.

FIGURE 2-10 The Azure AD Administration Portal

NEED MORE REVIEW? For complete information about the management of Azure AD, see *https://docs.microsoft.com/en-us/azure/active-directory-domain-services/.*

The following are core elements of Azure AD:

- **Azure AD Connect** This enables synchronization between an on-premises Windows AD and Azure AD. By implementing this, a company could enable all local users to access various Azure resources without adding them by hand and centralize membership administration.

- **Single sign-on (SSO)** This is a mechanism to identify users one time and allow them to access all AD services for which they have permission. It works using a centralized identification point for all related applications and services. When a user tries to use one of these applications, the application connects to a centralized point of authentication. The authentication center then asks the user for his or her credentials and authenticates the user (assuming the credentials are valid). If the user then attempts to access another application or service, that application or service connects to the authentication center, which grants access without requiring the user to provide his or her credentials again.

- **Federated single sign-on** This is an extension of SSO, which enables the identification of users outside the company domain by any other AD to allow them to access company resources. This is done through the Guest User link in the AD Administration Portal. This allows you to invite external users to access certain company resources, with permissions granted by the company's security administrators.

- **Multi-factor authentication (MFA)** MFA requires more than one mechanism to identify a user. It uses different user attributes, like what you know, what you are, and what you have. This can be a user choice, if enabled. With MFA, users must have more than one method of identification. One will be the user name and password, but the administrator could choose from various others, likes access code, device authorization, or image recognition. After this is configured, the next time the user accesses a resource, the centralized authentication site will ask the user which other mechanism (or mechanisms) he or she will use and launches a wizard to help the user configure it. Moving forward, each time the user authenticates, he or she will need to successfully navigate all the configured mechanisms.

> **NOTE** For more information about how AD works, see *https://docs.microsoft.com/en-us/azure/security/fundamentals/technical-capabilities#manage-and-control-identity-and-user-access*.

- **App registration** In some cases, you will need to authorize an application or a business consumer (instead of a user) without using personal credentials or sending user credentials to the storage service. For example, you might want to allow an external or anonymous user or a mobile or desktop application to retrieve information such as your product list without having to identify itself. In this case, you could use the Azure AD App Registration Portal. (See Figure 2-11.)

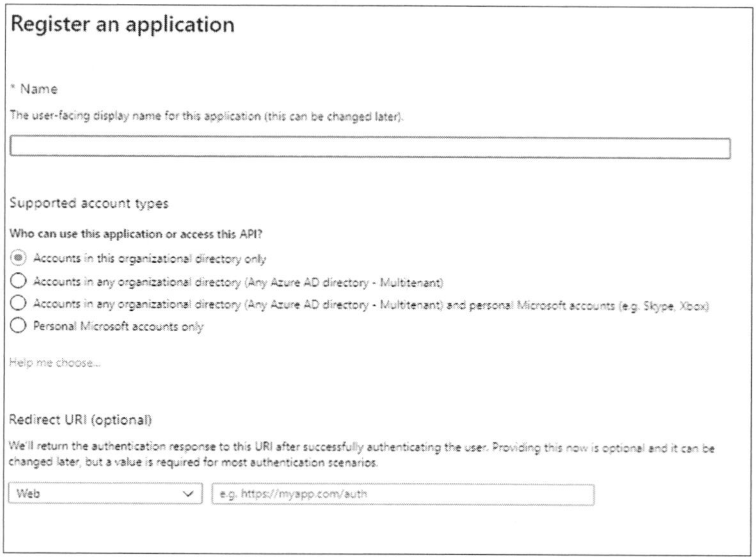

Register an application

* Name
The user-facing display name for this application (this can be changed later).

Supported account types

Who can use this application or access this API?

- ● Accounts in this organizational directory only
- ○ Accounts in any organizational directory (Any Azure AD directory - Multitenant)
- ○ Accounts in any organizational directory (Any Azure AD directory - Multitenant) and personal Microsoft accounts (e.g. Skype, Xbox)
- ○ Personal Microsoft accounts only

Help me choose...

Redirect URI (optional)

We'll return the authentication response to this URI after successfully authenticating the user. Providing this now is optional and it can be changed later, but a value is required for most authentication scenarios.

| Web | e.g. https://myapp.com/auth |

FIGURE 2-11 The Azure AD App Registration Portal

Figure 2-11 shows the information needed to register an app, such as a friendly name for user localization as well as authorization levels. The authorization level defines which users can authenticate themselves. For example:

- Only accounts in the organization's AD directory can use the application.
- Accounts in any Azure AD directory can use the app.
- Accounts in any Azure AD directory and personal Microsoft accounts (for example, Outlook.com, Live.com, or Hotmail accounts) can use the app
- Only personal Microsoft accounts can use the app.

    In addition, there is an optional redirect URI in case the application is a website.

    After you define the app, you can grant permissions to Azure storage. You can also choose to employ user credentials or to run as a background service or daemon without user credentials. To do so, you must create a client secret that the app can use to retrieve an access token when needed.

Role-based access control

Azure uses role-based access control (RBAC) to assign permissions to any resource. Roles map to one or more resource providers and establish the permissions for their assigned identities. Azure has more than 140 RBAC definitions. Some of these apply to any resource, while others are specific to a particular resource. To see the list for yourself, execute the following command in a command window or in Azure Portal Cloud Shell:

```
az role definition list -o table --query [].roleName
```

> **NOTE** Different subscriptions could have different RBAC definitions.

There are three basic roles for any resource:

- Owner
- Contributor
- Reader

These authorize and/or deny different actions against resources in general.

If you look at the definition of one of these roles, you can see how RBACs are defined. For example, you can execute the az role definition list -n Contributor command to see its definition, which is as follows:

```
[
  {
    "assignableScopes": [
      "/"
    ],
    "description": "Grants full access to manage all resources but does not allow you to
assign roles in Azure RBAC.",
```

```
    "id": "/subscriptions/<subscription>/providers/Microsoft.Authorization/roleDefinitions/
<RBAC>",
    "name": "b24988ac-6180-42a0-ab88-20f7382dd24c",
    "permissions": [
      {
        "actions": [
          "*"
        ],
        "dataActions": [],
        "notActions": [
          "Microsoft.Authorization/*/Delete",
          "Microsoft.Authorization/*/Write",
          "Microsoft.Authorization/elevateAccess/Action",
          "Microsoft.Blueprint/blueprintAssignments/write",
          "Microsoft.Blueprint/blueprintAssignments/delete"
        ],
        "notDataActions": []
      }
    ],
    "roleName": "Contributor",
    "roleType": "BuiltInRole",
    "type": "Microsoft.Authorization/roleDefinitions"
  }
]
```

In the permissions node, permissions are assigned to actions. The noActions node, however, lists actions that are *not* allowed for the role. So, in the case of the Reader role, only one action is permitted: the read action.

Notice that there are two other entries: dataActions and noDataActions. These apply specifically to users with permissions to manipulate data inside the resource. You can see this here in one of the specific roles for storage accounts, where all permissions are granted under dataActions:

```
    "permissions": [
      {
        "actions": [
          "Microsoft.Storage/storageAccounts/blobServices/containers/*",
          "Microsoft.Storage/storageAccounts/blobServices/generateUserDelegationKey/action"
        ],
        "dataActions": [
          "Microsoft.Storage/storageAccounts/blobServices/containers/blobs/*"
        ],
        "notActions": [],
        "notDataActions": []
      }
    ],
```

To handle permissions to access data or manage storage account resources, click the **Access Control (IAM)** option in the left toolbar of the Azure Portal storage account page. The Access Control page opens, displaying several tabs. In the first tab, Check Access, you can evaluate the roles assigned to yourself and, if you have security admin permissions, you can see the roles assigned to other users.

The Check Access tab also offers quick access to Add a Role Assignment, View Role Assignments, and View Deny Assignments buttons to enable you to manage the security of the storage account. For example, clicking the **Add a Role Assignment** button opens an Add Role Assignment panel, where you can specify the role to assign and to whom (an AD user, a service account, and so on). (See Figure 2-12.)

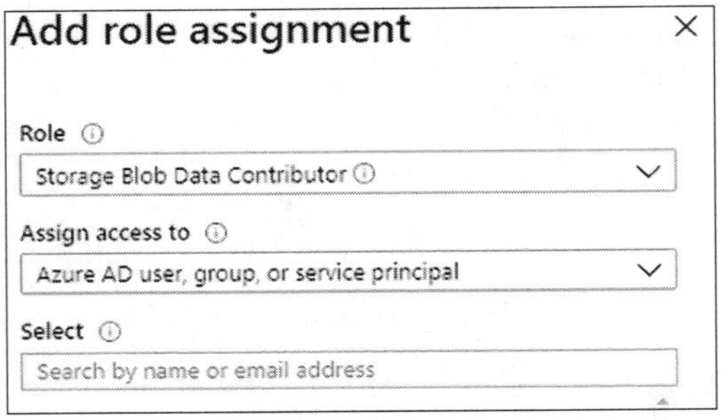

FIGURE 2-12 The Add Role Assignment panel

Other authorization methods

Some storage resources might use specific tokens or keys to allow connections. For example, Cosmos DB automatically generates two keys for writing and reading, and two more keys with read-only permissions.

For more fine-grained permissions, you can create users and assign them permissions for specific containers using code like the following sample in C#:

```
Database = cosmosClient.GetDatabase(databaseName);

User user = await database.CreateUserAsync(newUserName);
//assign write access permissions for the user at container level
Container container = cosmosClient.GetContainer(databaseName, collectionName);
await user.CreatePermissionAsync(
    new PermissionProperties(
        id: "writePermissionNewUser",
        permissionMode: PermissionMode.All,
        container: container,
        resourcePartitionKey: new PartitionKey(partitionKey)));
```

> **NEED MORE REVIEW?** For detailed information on other ways to create Cosmos DB permissions using Python, CLI, and API calls, see *https://docs.microsoft.com/en-us/rest/api/cosmos-db/create-a-permission*.

Storage accounts allow access using one of the automatically defined storage keys. You can also define more specific access by generating a shared access signature (SAS). When you generate a SAS, you get a specific URI pointing to the storage resource and a token with the specific identification and connection permissions. There are several ways to generate a SAS, including .NET libraries, PowerShell, Azure CLI, and the Azure Portal itself.

To generate a SAS in the Azure Portal, click the **Shared Access Signature** option in the left toolbar of the storage account **Overview** page. This opens a page where you can define the permissions for services and even a specific range of IP addresses. (See Figure 2-13.) Options on this screen are as follows:

- **Allowed Services** Define the services to which the SAS has permissions. These could be any of the following:
  - Blob
  - File
  - Queue
  - Table
- **Allowed Resource Types** Stretch the permissions inside the allowed service by specifying which kind of API access the SAS allows. For example, if you select Object, only the API part that manipulates specific objects can be used. Your options are as follows:
  - **Service** The entire service can be used.
  - **Container** Only the container can be used.
  - **Object** Permissions are granted to objects.
- **Allowed Permissions** Fine-tune the permissions enabled for the SAS, including the following:
  - Read
  - Write
  - Delete
  - List
  - Add
  - Create
  - Update
  - Process
- **Blob Versioning Permissions** Select the check box in this section to enable the deletion of versions of objects (when versioning is enabled).
- **Start and End Expiry Date/Time** These values must be entered, but you can define an end date far in the future.
- **Allowed IP Addresses** Identify an IP address or a range of IP addresses that are allowed to access the resource. Or leave it empty to allow any connection.

- **Allowed Protocols**   By default, only the HTTPS protocol is enabled, but you can enable the less-secure HTTP protocol as well.

- **Preferred Routing Tier**   Select the connection routing tier. The option is enabled if you published the endpoints in the firewall section.

- **Signing Key**   Choose which signing keys assigned to the account must be used. Remember, storage accounts automatically have two keys assigned, which you can change at any time.

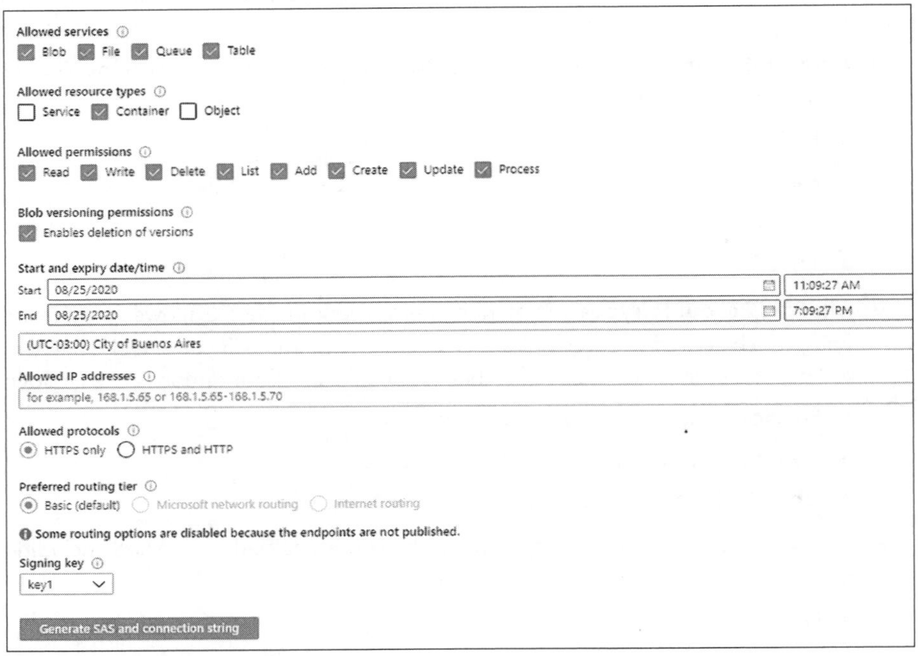

FIGURE 2-13   Defining a SAS

When you click the Generate SAS and Connections String button, you get the different connection strings for your selected services, each with the SAS assigned. You also get the SAS in a separate text box, which looks something like this:

```
?sv=2019-12-12&ss=b&srt=c&sp=rwdlacx&se=2120-08-25T03:00:00Z&st=2020-08-
25T03:00:00Z&spr=https&sig=K4kxlpdiU1muUCmKKtTzvl0KrUNc151pNlt%2F8Rnh8Ac%3D
```

Implement for high availability, disaster recovery, and global distribution

One of the most important benefits of using the Azure platform is its ability to immediately recover in the event of a disaster. Various components provide high availability and global distribution when needed. For example, by default, the system maintains three copies of any repository you deploy. Per the Azure service level agreement (SLA), this is the minimum number of copies to keep your data secure. However, the system can be configured to maintain even more copies, depending on your requirements. This applies to any storage — for example, disk images for virtual machines (which are in fact in blob storages). In this modern era, more and more copies are needed to distribute globally. This section reviews how to configure storage for this.

Storage account redundancy and high availability

When you create an Azure storage account, you must select one of the following redundancy options:

- **Locally redundant storage (LRS)**   Data is replicated on three disks within the same data center. If one of the disks fails, your data will remain intact on the other two. This is the lowest level of redundancy available. It's best for cases in which it is easy to refresh the original data or if you simply need to meet minimum data-governance standards. Still, it works fairly well — unless catastrophe strikes the data center. For example, if the data center experiences an earthquake, a fire, or some similar event, then the data on all three disks might be lost.

> **NOTE**   With LRS, the write process is managed in a synchronic way. So, the system notifies you of a successful write only after all three copies of the data have been updated.

- **Zone-redundant storage (ZRS)**   This replicates data in three different zones in the same region. Each zone is independent from the others at the hardware and base service levels (power, network, and so on). In the event of a failure, the system recovers automatically by adding a new zone replica. However, some delays may occur. As with LRS, the write process is synchronous for the three zones. For this reason, the write process must have resilient retry procedures. Similar to LRS, ZRS might be a good choice when you simply need to comply with data-governance restrictions.

> **NOTE**   Not all regions have this option available. See *https://docs.microsoft.com/ en-us/azure/storage/common/storage-redundancy#zone-redundant-storage* for an updated list of regions that support this option.

- **Geo-redundant storage (GRS)**   This option includes an additional LRS but in a different geographical region. The primary LRS is written synchronously and the data is sent asynchronously to the second LRS. The second LRS then repeats the synchronous process before confirming the second phase as committed.

- **Geo-zone-redundant storage (GZRS)**   This is identical to GRS except it involves a primary ZRS with a secondary LRS. At the time of this writing, this option was still in preview.

Here are some important considerations if you choose one of the geo-redundant options:

- You do not choose the second region. It is assigned automatically based on the primary region and is immutable.
- Updates to secondary regions occur asynchronously. If there is a failure across the entire primary region, part of the data would likely be lost because it takes some period of time for the second region to be updated. This is known as the *recovery point objective* (*RPO*). The RPO is usually less than 15 minutes, but a lot of data could be lost during that time. Of course, a total failure across an entire region rarely occurs, but it is something to consider.
- Read access from the secondary region is not enabled unless the user or the Azure team launches a recovery procedure. If you need read access at a secondary region, you must use alternate read-access geo-redundant storage (RA-GRS) or read-access geo-zone-redundant storage (RA-GZRS).
- Applications that perform reads from the secondary location must use a URL like *<your_account>-secondary.blob.core.windows.net*.

> **NOTE**   The last two geo-redundant options are not allowed for shared file storage.

> **NEED MORE REVIEW?**   To automatically switch between the primary and secondary read-only region, you can use a circuit-breaker pattern. A complete sample is available here: *https://github.com/Azure-Samples/storage-dotnet-circuit-breaker-ha-ra-grs*.

Storage account disaster recovery

High availability for storage accounts relies on procedural methods. There are no automatic backup procedures for storage accounts. However, you can enable soft deletes for blobs if the blob is not a data lake–enabled account. To do this, choose the **Data Protection** option (Tables, Blobs, Files, and so on) in the left panel of the Azure Portal service account page.

> **NOTE**   The soft delete option is not yet enabled for ADLS Gen2. However, there is a soft delete option for containers for this storage solution. However, you must request this from the Azure team.

When you enable soft deletes, you must specify a retention policy. This is seven days by default. In other words, you'll be able to see deleted items in the container for seven days. First, though, you need to enable the Show Deleted Blobs check box in Data Explorer.

Another manual protection could be enabling the Snapshots for Blobs setting in the Data Protection page. That way, you can take snapshots of any blob at any time. Snapshots are not performed automatically, however. You can take them manually in Data Explorer or programmatically using code.

NEED MORE REVIEW? To find out how to take snapshots programmatically and to view a code sample, see *https://docs.microsoft.com/en-us/azure/storage/blobs/snapshots-manage-dotnet*.

So, if ADLS Gen2 does not support automatic backup procedures or soft deletes, how can you protect data you have stored on this system from disaster? Fortunately, there are programmatic solutions to this problem, including PowerShell scripting, CLI batch commands, .NET libraries, and the REST API. Perhaps the most mature API is a .NET library for data movement: `Microsoft.Azure.Storage.DataMovement`.

NEED MORE REVIEW? For complete step-by-step instructions on using the DataMovement library, see *https://docs.microsoft.com/en-us/azure/storage/common/storage-use-data-movement-library*. The sample code at this link relates to creating a console application, but it is very simple to implement it as an Azure function triggered by a schedule.

Cosmos DB redundancy and global distribution

Cosmos DB is designed to be geographically distributed and uses redundancy and global distribution by default. In doing so, it automatically implements high availability, as the distribution makes it almost impossible for the service to become unavailable.

When you create a Cosmos DB account and enable geo-redundancy, you will have at least one other read-access replication in a geographically close region. (Refer to the practice exercise "Create a Cosmos DB account" earlier in this chapter.) To add more replicas to other regions, follow these steps:

1. Open *https://portal.azure.com* and use your credentials to log in.
2. Navigate to your Cosmos DB account.
3. In the left pane, under **Settings**, click **Replicate Data Globally**.

   The pane on the right displays a map showing your enabled regions. The primary region is dark blue with a check mark; the secondary region is lighter blue with a check mark. (See Figure 2-14.)

FIGURE 2-14   Cosmos DB geo-redundancy

4. Click any data center icon to add it as a replication zone.

   By default, Cosmos DB has just one write region and several read regions. If you need to implement multi-master distribution, change the **Multi-Region Writes** setting under **Configure Regions** to **Enable**.

   Each new region you add will now be write-enabled. Moreover, regions that can manage availability zones will now appear with a check box alongside them so you can enable that capability. This adds another layer of high availability to your implementation.

Using this graphical interface, you can very easily define your geo-replication preferences. The more regions you enable, the better your disaster-recovery profile. Of course, having more regions results in higher costs, as does enabling multi-region writes.

> **NOTE**   It is important to understand the cost-benefit effects of using different configurations with Cosmos DB accounts. To estimate costs, use the calculator on this page: *https:// azure.microsoft.com/en-us/pricing/details/cosmos-db/*.

Cosmos DB high availability

Several Cosmos DB configuration options enable you to define exactly what you want for your store. This is possible because the system maintains several copies of the same data.

As explained, a Cosmos DB account can contain one or more databases, and each database can contain several containers. Each container, which can store tables, graphs, and other collections, is stored in a physical partition in each region. And each partition has four copies of the same data. So, for example, if you define your Cosmos DB to use three different regions, then the system will maintain 12 copies of your data at any given time.

A single point of failure can occur if a database has just one write region and multiple read replicas and something bad happens, such as a misbehaving application or uncontrolled changes. To ensure the account reacts properly when this occurs, you can configure the **Automatic Failover** setting in the **Same Replicate Data Globally** page. (See Figure 2-15.) This same page also offers a **Manual Failover** button. You can click this to simulate a failover to see how the platform will fix the issue and how much time it will take.

FIGURE 2-15 Global replication and failover configuration

For true high availability, you must configure at least two regions for writes. If a failure occurs in one of these regions, the application will continue write activities in the other region if the Automatic Failover setting is enabled for the account. When this happens, Cosmos DB takes the following steps:

1. It automatically detects the failure in the primary region.
2. It temporarily promotes a secondary region to primary status to continue write operations.
3. When the original primary region comes back online, the system replicates changes in the temporary region to the original region.
4. It re-promotes the original region to primary status.

If the account has the Multi-Region Writes setting enabled, the system simply redirects the writes to another region in the event of a failure. This makes the system more responsive, with almost no delays.

You can also configure availability zones. Doing so creates replicas in different zones in the same region. When you define a multi-region write configuration, some locations will enable this option automatically. (At this time, not all regions support this.) This zone redundancy is available at no extra cost.

> **NOTE** Using these deployments, Azure ensures SLAs between 99.99% for single storage and 99.999% for geo-replication with multiple writes.

Implement relational data stores

Now it is time to look at relational data stores. You likely already know about and work with them, so you will not find details here about provisioning them in Azure. Instead, you will find important information about implementing a secure and reliable environment for your relational data.

This section discusses the implementation of relational data stores in Azure like the following:

- Azure SQL Database
- Azure SQL Managed Instance (SQL MI)
- Azure Synapse Analytics

The first part of this section includes practice exercises to perform deployments to create testing procedures.

Provide access to data to meet security requirements

As you have learned, Azure implements an entire platform to manage security and access to resources. This same platform applies to relational data stores. In this section you'll learn how to set up and secure an Azure SQL Database.

PRACTICE   Create an Azure SQL Database

To create an Azure SQL Database for use as a testing environment, follow these steps:

1. Open *https://portal.azure.com* and use your credentials to log in.
2. In the search box, type **SQL Databases**, and select the corresponding entry that appears in the drop-down list.
3. Click the **Add** button.
4. Open the **SQL Database Creation Parameters.docx** document in the **Templates** folder in the Chapter 2 companion content and use the values listed in the document to configure the SQL Database settings. Empty spaces are in some cells for you to complete.
5. Click the **Review + Create** tab. Then click the **Create** button.
6. Click the **Overview** link in the results page that appears.

> **NOTE**   Because you selected the Serverless option when configuring the database, the Auto-Pause feature was automatically enabled. This means the database will be stopped when not in use.

7. To prevent errors associated with the Auto-Pause feature, go to the database **Overview** page to automatically start the database.

If you are not using a fixed IP address, complete steps 9 through 11 to ensure your firewall will allow database access from your IP address. Otherwise, skip to step 12.

8. Click the server name link in the **Overview** page.

9. Click the **Show Firewall Settings** link.

10. Make sure your IP is in the list that appears. If not, click the **Add Client IP** button.

11. When the configuration, included your IP, is changed, click Save. (See Figure 2-16 and click **Save**.)

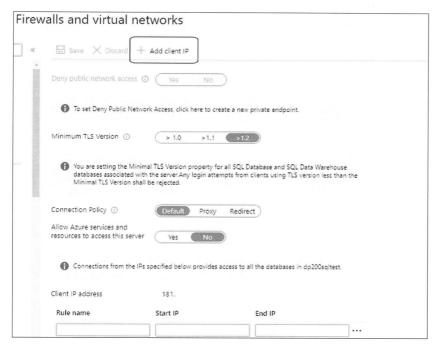

FIGURE 2-16 Enabling the client IP address in the SQL Server firewall

> **NOTE** In addition to the Web Data Explorer, which is part of the Azure Portal, there are several tools to connect and manage your Azure SQL Database. Here, you'll use Azure Data Explorer to follow the code and see the result for each sample. You'll find code samples and references to explanations in a notebook named SQL Database.ipynb, which is available from the companion content.

PRACTICE   Connect to your Azure SQL Database from Azure Data Studio

To test the code in the SQL Database.jpynb notebook, you need to establish a connection to your database. This practice exercise guides you through this process using Azure Data Studio.

1. Open Azure Data Studio.

> **NOTE** If you need to install Azure Data Studio, you can get the latest version here: *https://docs.microsoft.com/en-us/sql/azure-data-studio/download-azure-data-studio*.

2. Click the **Connections** button at the top of the left toolbar to open the Connections panel. (It has a server icon on it.) Alternatively, press **Ctrl+Shift+D**.

3. Click the **New Connection** button in the pane next to the toolbar.

   The Connection Details panel opens. It shows your existing connections (if any) and settings to create a new connection. (See Figure 2-17.)

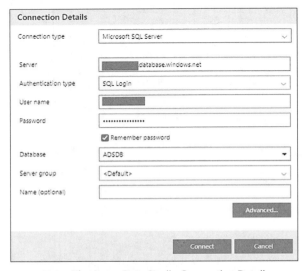

FIGURE 2-17 The Azure Data Studio Connection Details pane

4. Configure the following settings:

   - **Connection Type** Choose the type of database you are connecting to.
   - **Server** Type the name of your server. For an Azure SQL Database, this name must include the port, in the format <server_name>.database.windows.net,1433.
   - **Authentication Type** For this first connection, select **SQL Login**.
   - **User Name** Type the user name defined when creating the server.
   - **Password** Type the password defined when creating the server.
   - **Remember Password** Select this check box to configure the connection to remember the password.
   - **Database** Use the drop-down list to choose the database to which you want to connect. (This assumes the user name and password are correct; Data Studio will connect to the server to get the database name.)

- **Server Group** You can leave this set to the **<Default>** option, disable the setting, or opt to create a new server group to organize your connections.
- **Name** Type a friendly name for the connection.

> **NOTE** Clicking the Advanced button opens another tab with several options to fine-tune the connection, such as Connect Timeout, Application Intent, Security, and more.

5. Click the **Connect** button to establish the connection.

   After the connection is established, the Connections panel displays a Connections tree. It contains nodes for the different objects in the database. A panel with a database home page displays a list of the elements inside the database.

   Your next step is to open the notebook from the companion content and connect to your database.

6. In Azure Data Studio, in the **Connections** pane, click the connection to your database in the **Connections** tree to the left.

7. Press **Crtl+Shift+B** to see the Notebooks tree view.

8. Click the **Open Notebooks in Folder** button and select the folder from the companion content that ends with **Ch 2\Code\Data Studio**.

9. Click the **SQL Database notebook** in the tree.

   A notebook editor opens. (See Figure 2-18.)

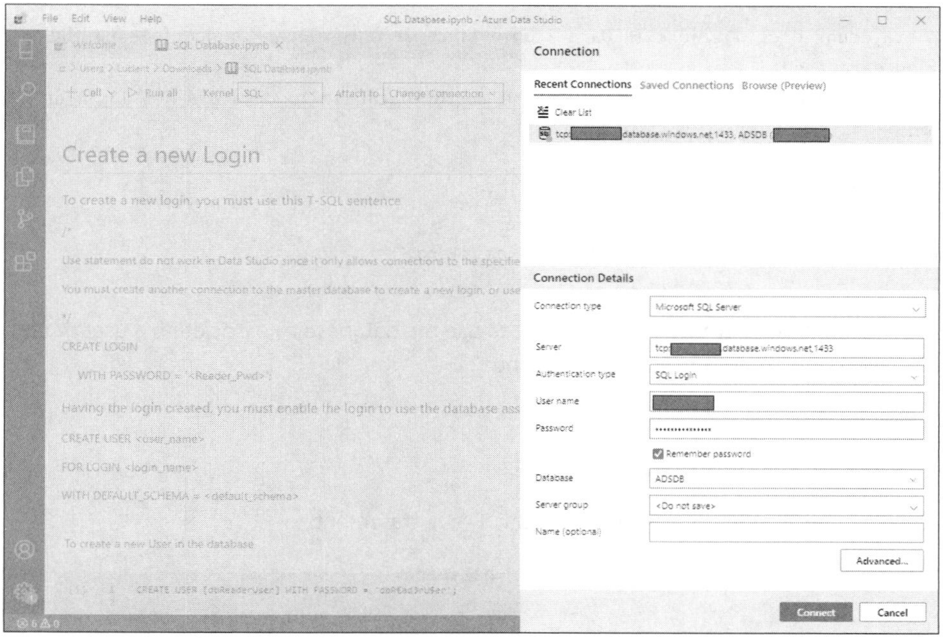

FIGURE 2-18 Link the notebook to a database connection

10. Open the **Attach To** drop-down list and choose **Change Connection**.

11. In the **Recent Connections** list, click the connection to your database.

12. Click the **Connect** button.

13. Open the **Attach To** drop-down list again. It will show the name of your connection.

User name and password

Entering a user name and password is the primary way to access your database, and at first, it will be the method you have to use. However, it is the riskiest method, because the user name defined when the database was created has no restrictions.

In the previous practice exercise, you used a user name and password to establish a connection to your database. Now, it is time to create other accounts to avoid using the administrator account for everything.

There are two different approaches to creating accounts:

- **Login** The account is stored at the server level, which means it is part of the master database content — that is, the database where the different elements for the server are stored. After the account is created, it is linked to roles in the databases to which you want to allow access, as you will see shortly.

- **User** The account is defined and stored in the database where it is created and can be used only for that database. As a login, the user account could be linked to database roles.

To create a new login, use this T-SQL statement:

```
/*
    Use statement does not work in Data Studio since it only allows connections to the
specified database
    You must create another connection to the master database to create a new login or
use SSMS.
*/
USE master;
GO
CREATE LOGIN <ReaderUser> WITH PASSWORD = '<password>';
GO
```

After creating the login, you must enable it to use the database by associating a user name to the login account. Use code like the following:

```
CREATE USER <user_name>
    FOR LOGIN <login_name>
    WITH DEFAULT_SCHEMA = <default_schema>
```

To create a new user in the database, use this T-SQL statement:

```
CREATE USER <dbReaderUser> WITH PASSWORD = '<password>';
```

Having the user defined in the database or at the server level, you must assign the account to one or more database roles. Each database has a set of roles defined by default. You can use the following statement to see the defined roles in your database:

```
SELECT DISTINCT
    [name] AS [UserName]
  FROM
    [sys].[database_principals] AS [pr]
  WHERE [type_desc] = N'DATABASE_ROLE'
  ORDER BY
      [UserName];
```

Table 2-7 details the roles created by default in a new database.

TABLE 2-7 Database roles

Role Name	Permission Granted or Denied
db_accessadmin	Managing access permissions
db_backupoperator	Reading physical data to perform backups and restores with no access to the information stored inside the database
db_datareader	Reading data
db_datawriter	Inserting and updating information
db_ddladmin	Creating, modifying, and dropping objects from the database, like tables, indexes, views, and so on
db_denydatareader	Denying read access
db_denydatawriter	Denying write access
db_owner	Performing any action in the database
db_securityadmin	Creating, modifying, and dropping accounts and managing access permissions
public	Reading and writing data but not managing objects, permissions, or accounts

To assign a user to a role, use a T-SQL statement like this:

```
EXEC sp_addrolemember N'<role_name>', N'<user_name>'
```

To see which users are assigned to which roles, you can use a statement like the following:

```
SELECT DISTINCT
    [pr].[name] AS [User],
    [pr].[authentication_type_desc],
    [Roles].[name] AS [Role]
  FROM
    [sys].[database_principals] AS [pr]
        INNER JOIN
          [sys].[database_role_members] AS [rm]
        ON [pr].[principal_id] = [rm].[member_principal_id]
        INNER JOIN
          [sys].[database_principals] AS [Roles]
        ON [rm].[role_principal_id] = [Roles].[principal_id]
  WHERE [pr].[type_desc] <> N'DATABASE_ROLE';
```

Active Directory authentication

It is a best practice to manage users in a centralized place for any enterprise access. This requires you to enable AD authentication for your Azure SQL Database. When AD is enabled for your database, users can authenticate themselves using three methods:

- **Active Directory – Universal with MFA** Use this when you require multifactor authentication for the user.
- **Active Directory – Password** Use this to force the user to enter an AD login name and password.
- **Active Directory – Integrated** With this, the user employs the security identification (SID) assigned by the current user logged to the client computer.

When you use integrated security with on-premises implementations, the server running SQL Server is linked to the authentication domain. In this case, the Azure SQL Database runs in some virtualized environment outside your domain.

To obtain information about the account it is trying to authenticate, the Azure SQL Database needs to check it against AD. Figure 2-19 contains a schematic representation of how Azure SQL Database authenticates AD users, described by the following steps.

1. A user tries to connect to the database by any of the AD-related mechanisms (Active Directory – Universal with MFA, Active Directory – Password, or Active Directory – Integrated).
2. The database engine connects to AD using an account defined specifically for the SQL service.
3. Using this account, the SQL service queries the AD for the user SIDs. These include the user SID and all SIDs from the groups to which the user belongs.
4. The SQL engine locates authorized accounts in the database for the supplied SIDs.
5. Using the role assignment associated with the SIDs, the engine gives the connection the appropriate permissions.
6. Based on these permissions, the engine executes the user call.

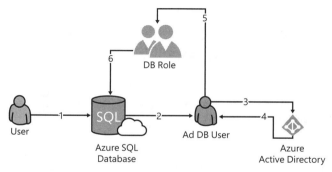

FIGURE 2-19 AD user authentication in Azure SQL Database

Based on this functionality, you must have an AD account with the required permissions to query the AD for the user identification and membership. On the other hand, the Azure SQL Database must receive from the user the different AD authentication methods. The steps in Exercise 2-5 and Exercise 2-6 are used to enable and configure AD authentication.

EXERCISE 2-5   Establish an AD admin account

This exercise contains the steps to add an AD account to Azure SQL Server to enable it to query AD for authorized accounts.

1. In the Azure Portal, on the database **Overview** page, click the **server name**.
2. In the **Settings** group in the left pane, click the **Active Directory Admin** option.

   This will enable you to define one AD administrator to control logins for other users in the future. This account must query AD for validation and group memberships.
3. Click the **Set Admin** button.
4. Select the user you want to assign as SQL Server administrator for AD authorizations.

   > **NOTE**   You must select an account that does not belong to a Microsoft public domain like outlook.com, live.com, or Hotmail.com, or to onmicrosoft.com domains.

5. Click the **Save** button.

EXERCISE 2-6   Add user or group accounts

This exercise demonstrates the procedure to add AD users or groups as database-enabled accounts.

1. Connect to the database using the account you recently assigned as admin.

   For SSMS, you can use the Azure Directory – Password authentication method. For Azure Data Studio, you can define a new connection using the Azure Active Directory – Universal with MFA Support method.
2. Select your database.
3. Click the **New Query** button in the toolbar.

4. Type the following statement to create a new contained user (not a server user, but a user just for this database, contained in it) for the database:

```
CREATE USER [<Azure_AD_principal_name>] FROM EXTERNAL PROVIDER;
```

> **NOTE** Since the @ symbol is used for variable identifiers in SQL Server, you must delimit the user name with brackets.

> **NOTE** You can use the same syntax to add an AD group to enable access to all its members.

5. To add an AD user to a database role, use the following syntax:

```
EXEC sp_addrolemember N'db_datareader', N' user@domain '
```

Protecting the connection

Another important security matter pertains to the network connection. To manage this, each SQL Server uses a firewall to block unauthorized connections — which are *all* connections when the server is originally deployed. You can manage the firewall behavior and authorize certain IP addresses to access the server by following the procedure in Exercise 2-7.

> **NOTE** In the server creation wizard, you can leave the client IP for the deployer enabled so you have another option to allow connections from other resources in the Azure platform.

EXERCISE 2-7   Manage the firewall configuration

This exercise guides you through the steps to change firewall settings for accessing a database server.

1. In the Azure Portal, on the database **Overview** page, click the **server name**.
2. To the right of the Overview panel, below the **Server Admin** name, click the **Show Firewall Settings** link.
3. To establish a level for TLS encryption, click the **1.2** button.
4. Under the **Connection** policy title, set the **Allow Azure Services and Resources to Access This Server** option to **On**.

   You'll see a list of the IP addresses authorized to access the server. You can remove any of these or add new ones. At the top of the panel is a button to add your IP address. Finally, you can add virtual private networks (VPNs) to allow connections to access the server from inside the VPN.

   > **NOTE** Later in this chapter, in the "Manage data security" section, you will see other ways to further secure your data.

Implement for high availability and disaster recovery

The Azure services contract establishes strict SLAs for all services provided. Most of them start with a 99.99% (so-called "four-nines") level SLA. These SLAs cover the infrastructure as a service (IaaS) for the physical buildings, power supply, networking, and hardware. For services, there are SLAs for the platform as a service (PaaS) as well.

To enable you to select the best cost-benefit-availability ratio, services are organized in service tiers. At the same time, each service tier can be configured for zone redundancy to enhance data availability. Table 2-8 compares the different service tiers and SLAs.

TABLE 2-8   SLAs for service tiers

Service Tier	Service Level Agreement
General Purpose/Standard	99.99%
General Purpose/Standard with zone redundancy	99.995%
Business Critical/Premium	99.99%
Business Critical/Premium with zone redundancy	99.995%
Hyperscale without replica	99.9%
Hyperscale with one replica	99.99%

> **NOTE**   Part of high availability relates to how you establish the storage and redundancy for your SQL Database instances.

General Purpose/Standard

The General Purpose/Standard tier has two names because Azure SQL Database can be measured in vCores or in database transaction units (DTUs). The General Purpose name pertains to vCore, while the Standard name applies to DTUs.

> **NOTE**   For more information on the different ways to measure resource utilization levels, costs, and features, see *https://docs.microsoft.com/en-us/azure/azure-sql/database/service-tiers-general-purpose-business-critical*.

This tier separates compute services from storage services to protect your data. Non-critical elements like processing, cache, and temporary storage are managed differently from your real data.

Processing, cache, and temporary storage are managed by Azure Service Fabric. It runs and controls the SQL service, the resources needed to process your work, and the volatile storage used by it. It also controls the health of the compute node and keeps copies of it on hold in case something goes wrong with the primary one.

Meanwhile, this tier uses redundancy from Azure storage to keep your data secure and protected. So, at any time, there are at least three copies of your data in the same data center.

Moreover, when you configure this service tier for a database, you can enable zone redundancy, which matches the zone redundancy of the underlying Azure storage (described in the first part of this chapter).

Figure 2-20 shows a schematic representation of the components of the General Purpose/Standard tier, where the compute service inside the Azure Service Fabric rack interacts with an Azure storage account with three replicas. A separate box signifies a fourth storage, representing the zone-redundancy option.

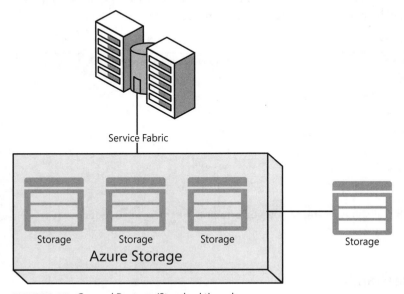

Service Fabric

Storage   Storage   Storage   Storage

Azure Storage

FIGURE 2-20   General Purpose/Standard tier schema

Business Critical/Premium

This tier uses a different approach to implement high availability. The compute and storage services exist as a unit but are redundant. Based on the implementation of availability groups for on-premises SQL Server, this tier has four cluster nodes in sync. One node acts as the primary and sends push requests to the other three nodes to keep data updated.

Having three replicas, this tier uses a read scale out procedure to redirect read activities (for example, for analytical processing) to secondary nodes. This reduces the load on the main node and improves its response times. The tier also automatically switches to another node if a failure occurs on the primary node. A new node is then created to replace the damaged one.

In highly intensive update workloads, this tier's response time is better than the General Purpose/Standard one. At the same time, it is more reliable; having four nodes makes it very difficult to lose service.

Another protection method for this tier is geo-redundancy, which enables you to maintain nodes in different regions. That way, if all the nodes in one region go down — perhaps due to a catastrophic event like an earthquake or hurricane — the system can divert to nodes in a different region. This feature is available for all database levels.

Because replicas are in different zones, a secondary replica could be used as a primary source for another replica, creating a sort of chain. Of course, there will be some delays when updating the third replica, but in extreme, high-disaster recovery situations, this approach works well.

PRACTICE   Enable geo-replication

To enable geo-replication for a database, follow these steps:

1. Open *https://portal.azure.com* and use your credentials to log in.

2. Select your database to see its **Overview** page.

3. In the **Settings** group in the left pane, click the **Geo-Replication** option.

   A **Geo-Replication** page opens that displays your current location and the zones in which you can establish a replica.

4. Select the desired location for the replica.

5. A **Create Secondary** page opens. Configure the following settings:

   - **Secondary Type**   If you are defining a replica for a serverless database, this setting fixed as **Readable**.

   - **Target Server**   You will probably need to create a new server in the destination region.

   - **Pricing Tier**   The default value will match the tier of the target server you selected.

6. Click the **OK** button.

Implement data distribution and partitions for Azure Synapse Analytics

Azure Synapse Analytics uses a scale-out model. This means it can add more and more compute resources on demand. Information is stored in Azure storage from a variety of sources, in different formats, comprising a big data store. ML algorithms and other data-analysis technologies like Hadoop, Spark, and so on process and train the data to gain insights. Finally, the information is stored in Synapse tables (called *pool tables*).

Scaling Azure Synapse Analytics

Consolidation, analysis, aggregation, and queries are performed using a set of compute nodes. When a user makes a query, the query accesses the control node. This node, which uses a massive parallel processing (MPP) engine to prepare the query for parallel processing, sends it to compute nodes. Meanwhile, other techniques work in the background. Figure 2-21 shows a schema of the MPP structure.

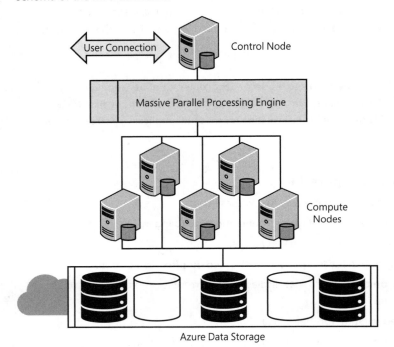

FIGURE 2-21 MPP architecture

Because several resources are involved, including compute nodes, storage, and other services, when you need Azure Synapse Analytics, you purchase a pool. The pool is a set of resources acting in sync to execute any action required over the data.

Data distribution

Due to the huge volume of data in analytical workloads, Azure Synapse Analytics "divides and conquers" to accelerate response times. That is, it segments data into parts and distributes it across Azure storage. Each segment is then used by a different compute node to perform parallel queries.

> **NOTE** A query can be divided into as many as 60 parallel queries, one for each distribution.

You can choose the distribution pattern of the data. The options are as follows:

- **Hash** This slices the rows into several distributions, one row per distribution. Consider using this option when you have a very large amount of fact data (more than 2 GB) in a star schema or when you know the data in a fact table will likely be subject to frequent changes.

- **Round robin** The default option, this slices an entire table into chunks to be distributed to several nodes. This occurs before query execution, which means it might be some time before query processing begins. You use this option if you have not identified a key column, which could be used to distribute a hash. Another scenario is when the data is recent and needs to be processed so you can gain some insights into it and how it might be better distributed.

- **Replicated tables** This maintains a copy of the table for each compute node. A parallel query is defined to obtain only a segment of rows in each compute node. This option works very well for small tables that can be quickly replicated through the nodes and analyzed in parallel. However, there is a cost involved in the replication process, so this is not a good option for large tables.

> **NOTE** To maintain data consistency, Azure Synapse Analytics calls Data Movement Service (DMS) any time a process starts to ensure that all the data is available for all the nodes.

Table partitions

Another way to organize your data in Azure Synapse Analytics is by defining table partitions. This is not the same as in OLTP SQL Server because in this case, the partition determines how the data is distributed and the syntax is different. It does not replace data distribution, but rather complements it, because it works with any of the described distributions.

There are a few considerations with table partitions:

- Table partitions apply only on one column. There is no way to define multiple partitions on different columns or to define partitions by combining columns.

- There is no benefit to defining partitions using a column with very different values because the partitions become small and consume more resources.

- There are no standard low or high limits for the size of each partition, but they must match the utilization.

One good candidate for partitions is a column containing dates. By slicing data by date groups, you can better manage and query it. Usually, updates of this data will be for recent dates — probably with more inserts than other updates. At the same time, queries of the data will probably filter and/or sort by date. And, if you want to remove data, it will probably be the oldest data. This could mean removing one or more partitions — an action that takes less time than deleting the rows one by one.

> **NOTE** Once defined, a good review of resource consumption and response times is generally the best way to establish which partition schema is best.

To create a table with a partition by date, you can use a statement like the following one:

```
CREATE TABLE [dbo].[FactSales]
(
      [ProductKey]          int            NOT NULL
,     [OrderDateKey]        int            NOT NULL
,     [CustomerKey]         int            NOT NULL
,     [PromotionKey]        int            NOT NULL
,     [SalesOrderNumber]    nvarchar(20)   NOT NULL
,     [OrderQuantity]       smallint       NOT NULL
,     [UnitPrice]           money          NOT NULL
,     [SalesAmount]         money          NOT NULL
,     [Discount]            real           NOT NULL
)
WITH
(    CLUSTERED COLUMNSTORE INDEX
,    DISTRIBUTION = HASH([ProductKey])
,    PARTITION    (    [OrderDateKey] RANGE RIGHT FOR VALUES
                       (20030101,20040101,20050101,20060101,
                       20070101,20080101,20090101,20100101,
                       20110101,20120101,20130101,20140101,
                       20150101,20160101,20170101,20180101,
                       20190101,20200101,20210101,20220101
                       )
                 )
)
```

> **NEED MORE REVIEW?** For detailed information about managing table partitions, see the following: *https://docs.microsoft.com/en-us/azure/synapse-analytics/sql-data-warehouse/sql-data-warehouse-tables-partition.*

Implement PolyBase

PolyBase is an Azure SQL Database feature that enables you to refer to external data as though it is a table inside a database and to use it in your queries and joins. At this time, you can link external data from:

- Azure storage accounts
- Oracle Server
- SQL Server (on-premises or in the cloud)
- Teradata
- MongoDB

In the same way that Azure SQL Server must have a way to connect to Azure AD to authorize users, it will also need credentials to connect to any external services. Those credentials cannot be stored as plain text. Some sort of encryption plan is required.

In Azure SQL Database, all you need to do is to create a master key, which will be used by SQL Server to encrypt and store security information. You can use this T-SQL statement to create the master key:

```
CREATE MASTER KEY ENCRYPTION BY PASSWORD = '<strong_password_here>';
```

Having the master key, you can define the credentials for any service you want to link to. For example, if you want to query data from Azure Blob storage, you do the following:

1. Add a user with one of the storage keys from the storage accounts:

   ```
   CREATE DATABASE SCOPED CREDENTIAL AzStorageCred
   WITH IDENTITY = 'user', Secret = '<azure_storage_account_key>';
   ```

2. Define the external source pointing to the Blob storage URL:

   ```
   CREATE EXTERNAL DATA SOURCE <DataSourceName> WITH (
    TYPE = HADOOP,
    LOCATION = 'wasbs://<container_name_here>@<storage_account_name_here>.blob.core.
    windows.net',
    CREDENTIAL = AzStorageCred
   );
   ```

   Notice it uses HADOOP as type, since it is the API used to get data from blob storage/ Data Lake.

3. Define which type of file are you going to get. For example, this statement defines a CSV file format:

   ```
   CREATE EXTERNAL FILE FORMAT CsvFileFormat WITH (
    FORMAT_TYPE = DELIMITEDTEXT,
    FORMAT_OPTIONS (FIELD_TERMINATOR =',', STRING_DELIMITER = '0x22',
    FIRST_ROW = 2, USE_TYPE_DEFAULT = TRUE));
   ```

   The file types could be any of the following:

   - **Optimized Row Columnar (ORC)**   This requires Hive version 0.11 or higher.
   - **Record Columnar File (RCFile)**   This needs to specify a Hive Serializer.
   - **DelimitedText**   This specifies a text format with column delimiters.
   - **JSON**   This is applicable to Azure SQL Edge only.

NEED MORE REVIEW? For detailed and updated information about PolyBase connections, see *https://docs.microsoft.com/en-us/sql/t-sql/statements/create-external-data-source-transact-sql.*

4. Create the table definition, matching the content you want to get:

```
CREATE EXTERNAL TABLE  Countries_Data
( [CountryCode]                  NVARCHAR(3),
  [continent]                    NVARCHAR(100) NOT NULL,
  [location]                     NVARCHAR(100) NULL,
  [population]                   INT NULL,
  [population_density]           NUMERIC(18, 5) NULL,
  [median_age]                   NUMERIC(18, 5) NULL,
  [gdp_per_capita]               NUMERIC(18, 5) NULL,
  [hospital_beds_per_thousand]   NUMERIC(18, 5) NULL,
  [life_expectancy]              NUMERIC(18, 5) NULL
)
WITH (
 LOCATION = '/<Path_To_The_Blob_File>/',
 DATA_SOURCE = <DataSourceName>,
 FILE_FORMAT = CsvFileFormat
);
```

Manage data security

This chapter reviewed several ways to manage access to your data. It also analyzed how to configure your data storage for high availability, including procedures for disaster recovery. This section discusses how you can protect your data "all the way" by using data masking and encrypting data at rest and in motion.

Implement dynamic data masking

Even if you've secured all your resources and identified all access vectors to your data, there is still a chance that someone could try to access data they are not allowed to see. We are not referring to people who breach your system to gain access, but rather to those who already have access to some amount of data.

NOTE You do not want all data to be accessible by all users.

In these situations, dynamic data masking can help you protect your data. Dynamic data masking enables you to configure certain columns in your tables to prevent their information from being displayed to non-authorized users. It is different from using features like views to filter which columns the user can see or establishing column-level permissions. That is, you do not deny access to all the data in a column; you just make it so the user can see only part of it.

You might use dynamic data masking if, for example, you wanted to prevent an unauthorized user from viewing the email addresses of all users, but you want that information to be partially displayed for verification purposes in email communications — as with messages that read something like this: "you are subscribed to this information with your xxxxx@----.com address." Here, part of the information hidden.

Another example is when information about payment methods, such as credit cards, needs to be verified but not with the *full* information. For example, suppose a customer-service representative is calling a buyer to inform her that a delivery could be delayed. The representative can prove his identity to the buyer by providing info about the credit card that was used, but he will not be allowed to see the credit card info in its entirety.

You can define a mask for a column in the table definition DDL or by altering the table after you create it to apply any of the masks detailed in Table 2-9.

TABLE 2-9 Dynamic Data Masking types

Data mask type	How it works	Syntax
default()	Data is masked based on its data type, strings are masked by four or fewer *X* characters, numbers are masked as 0 values, and dates are masked using a fixed date (the first day of 1900).	MASKED WITH (FUNCTION = 'default()'
email()	A masking pattern is used to hide the user email and the domain where the email comes from with a format like the following: <First_Letter>XXX@XXXX.com.	MASKED WITH (FUNCTION = 'email()'
random()	This mask applies to numeric data types and generates random values for each row from the range specified in the definition statement.	MASKED WITH (FUNCTION = 'random(<start range>, <end range>)'
Custom string	With this mask type, you indicate how the mask must be generated by defining the prefix and suffix characters displayed and leaving the values in the middle masked. In the code sample in the cell to the right, a mask for a credit card number is defined, allowing the user to just see the last four numbers.	MASKED WITH (FUNCTION = 'partial (0,"XXXXXXXXXXXXXXX",4)'

Dynamic data masking applies only to users who have select permissions — for example, like members of the db_datareader role, but not of the db_datawriter or db_owner roles. You can configure an account to allow it to see unmasked data by granting the UNMASK permission.

Of course, for any operation that involves obtaining information from masked columns, that information will always be masked — even when it is used to construct other objects such as table expressions, temporary tables, and so on. If the user has been granted access to masked data, he or she will always see that masked data, no matter how he or she uses it.

PRACTICE   Mask a column using the Data Query Editor

In this practice exercise, you will modify a column in a table in the sample database to see the effects of dynamic data masking. You will complete this practice using the Query Editor in Azure Portal. Follow these steps:

1. Open *https://portal.azure.com* and use your credentials to log in.

2. Select your database to see its **Overview** page.

3. In the **Data Query Editor** in the left pane, type the following statement to modify the Customer table to mask the email column:

```
ALTER TABLE [SalesLT].[Customer] ALTER COLUMN EmailAddress
ADD MASKED WITH (FUNCTION = 'email()')
```

4. Perform a select from the table to see the result:

```
SELECT CompanyName, EmailAddress FROM [SalesLT].[Customer]
```

Because you are connected as db_owner, you will see the information in its entirety.

5. To test the dynamic data mask, create a temporary user and grant it select permissions for the Customer table like so:

```
CREATE USER [SelectOnlyUser] WITHOUT LOGIN;
GRANT SELECT ON [SalesLT].[Customer] TO SelectOnlyUser;
EXECUTE AS USER = 'SelectOnlyUser';
```

6. Perform a select, as in the following statement:

```
SELECT
        [CompanyName],
        [EmailAddress]
    FROM
        [SalesLT].[Customer];
```

7. Revert and drop the test user like so:

```
REVERT;
DROP USER
    [SelectOnlyUser];
```

PRACTICE   Mask a column using the Dynamic Data Masking option

If you prefer not to use code, you can modify the masked elements in your database using the Dynamic Data Masking option in the Azure Portal. Follow these steps:

1. Open *https://portal.azure.com* and use your credentials to log in.

2. Select your database to see its **Overview** page.

3. Click the **Dynamic Data Masking** option in the **Security** group in the left pane.

   An **Add Masking Rule** pane opens. (See Figure 2-22.)

FIGURE 2-22 Adding a dynamic data mask to a field

4. Configure the following settings:

- **Schema**   Choose the schema to mask.
- **Table**   Choose the table to which you want to apply the mask — in this case, **Customer**.
- **Column**   Choose the column to which the mask should apply.
- **Masking Field Format**   Choose a predefined mask format — for example, **Credit Card Value**, **Email**, and so on.

5. Click **Add**.

> **NOTE**   Predefined masks are not available in T-SQL code.

Encrypt data at rest and in motion

To secure your data, Azure uses different encryption patterns for when it is in storage (at rest) and during data transmission (in motion).

Encrypt data at rest

Data stored inside an Azure storage account is automatically encrypted. This includes data in disk storage for SQL Database and other database technologies implemented on virtual machines. This data is said to be at rest. Encrypting data at rest is required by law in several countries, including the United States, as outlined in *US Federal Information Processing Standard (FIPS) Publication 140-2*.

NOTE One way to encrypt data at rest is by restricting physical access to the storage media — for example, by removing a hard drive from the data center.

The encryption process for data at rest is implemented in two layers. One encryption key is used to encrypt the data on the physical media. A second master encryption key is used to encrypt the data encryption keys.

This method uses different storage combinations for these keys:

- **Service-managed keys** All key management is performed by Azure services, including the generation and storage of the encryption keys.

- **Customer-managed keys** The encryption process is handled by Azure services but keys are created by the customer and stored in Azure Key Vault.

- **Customer-managed keys on customer-controlled hardware** This is similar to customer-managed keys, but the keys are stored in an on-premises data center under the control of the customer. This method is more complex to set up and maintain, but there could be specific cases requiring this level of security isolation.

You need not configure anything to encrypt your data. Azure storage accounts are encrypted automatically, and all content in every format is encrypted by default. You can, however, select the encryption mode. To do so, select the **Encryption** option in the storage account settings and specify whether you prefer to use service-managed keys or customer-managed keys. If you choose to use customer-managed keys, you will be prompted to provide the storage and the key. This can be the name of an Azure Key Vault or a URL to the key. (Note that if you go the Azure Key Vault route, you can choose whether to upload an existing Key Vault or create a new one.)

NOTE The Azure Key Vault is unique by subscription and is not part of any resource group.

Similarly, all your SQL Databases, Synapse databases, and SQL MIs have Transparent Data Encryption (TDE) enabled by default. To view (and if you want, disable) TDE, choose the **Transparent Data Encryption** option in the **Security** group in the left pane of your database's Azure Portal page.

Cosmos DB storage is automatically encrypted as well. However, it cannot be changed. Only when you deploy a new Cosmos DB account can you choose between service-managed and customer-managed keys, in which case you must provide the URI to the key stored in Azure Key Vault.

There is a special case for data encryption at rest: when the customer chooses not to use Azure platform encryption and instead employs custom client application encryption. This is possible using any encryption method. In this scenario, data is simply sent to storage already encrypted.

Encrypt data in motion

Encrypting data in motion means keeping data encrypted while it is being sent to or received from a client application. For example, in Azure SQL Database, all communication between the server and client application is established using Transport Layer Security (TLS).

You cannot choose a different method to secure data in motion with Azure SQL Database. It always uses TLS. However, you can configure which version of TLS is used.

In general, it's best to use the most recent version, because it results in better communication security. In some special cases, though — for example, with legacy applications — you might need to use an older version. (Still, you should try to use as recent a version as you can.)

To change the TLS level, follow these steps:

1. Open *https://portal.azure.com* and use your credentials to log in.
2. Navigate to your SQL Server.

> **NOTE** This is a per-server configuration, not a per-database one.

3. In the left pane, under **Security**, click the **Firewall and Virtual Networks** option.
4. For the **Minimum TLS Version** setting, choose the minimum TLS version required.

Azure storage accounts also use secure transfer, including TLS. Storage accounts with secure transfer enabled will reject access from any source not using the HTTPS protocol.

> **NOTE** Secure transfer does not work if you use a custom domain name. In that case, the option will be disabled.

For Cosmos DB accounts, TLS 1.2 is used by default in new accounts. To change this, you must request it from service support for the specific account.

For all internal communication between data centers, zones, and regions, Azure infrastructure uses IEEE 802.1AE MAC Security Standards (MACsec). This is managed by the platform, with no customer intervention.

Summary

- Cosmos DB can store different kinds of data, just by defining which API to use. Internally, it stores all data in small atom-record-sequence (ARS).

- Some APIs for Cosmos DB are designed to facilitate "shift and lift" from on-premises storages.

- Cosmos DB usually indexes all content to facilitate queries and searches.

- A Cosmos DB account can contain more than one database, which can have several containers to store data in them.

- The API is defined by account, but you can create as many as 100 accounts per subscription.

- Cosmos DB has programmatic capabilities through the implementation of triggers, procedures, and functions, in JScript language.

- Cosmos DB enables geo-replication with very fast update responses. Any region can be updated and replicated to the others.

- You can use Cosmos DB as a data source for Azure Synapse Analytics by enabling column stores.

- Cosmos DB is primarily a document storage with extended capabilities.

- The logical partitions for Cosmos DB do not determine how the data is physically stored, but how it is distributed from a query point of view.

- Azure Table storage has moved to Cosmos DB for better performance and search capabilities.

- Being distributed in nature, Cosmos DB requires a level of data consistency according to the data needs.

- You implement a data lake by using an Azure Data Lake Storage (ADLS) Gen2 account with hierarchical namespace enabled.

- ADLS can be accessed by several Azure services and by client applications as well.

- ADLS data is stored in Azure Blob storage repositories, which can be accessed by different APIs like NDFS, ABFS, and REST API.

- It is not recommended to use ADLS Gen1 because it will be completely deprecated soon.

- Azure storage preserves information using local, zone, and geo-replication, with at least four local copies.

- Azure access authentication is managed by each resource with custom login methods like user names and passwords or connection keys.

- Azure authentication can be managed by Azure Active Directory (AD) to allow access to resources. Each service establishes the mapping between the resource and the AD account internally.

- Azure resource permissions must not be assigned individually. Rather, they must be assigned using role-based access controls (RBACs).

- You can assign RBACs to applications accessing resources by registering them in Azure AD.

- All Azure resources have features to manage high availability, disaster recovery, and global distribution.

- Some level of disaster recovery — at the very least, having several copies of the same data in the same data center — is available on almost all service levels except those specifically designed for testing purposes.

- Disaster recovery features support high availability because they enable the Azure platform to automatically switch to a copy when the master process fails.

- Zone redundancy, when available, ensures a quick response to failure by keeping copies in nearby regions.

- Backup procedures are implemented by default and can be customized.

- Cosmos DB geographically distributes data storage automatically to keep the data close to the consumer.

- Azure SQL Database and Azure Synapse Analytics use similar functionality to protect your data and make sure it is always available, keeping copies and/or twin server instances, depending on the service level.

- Azure SQL Database and Azure Synapse Analytics can use the PolyBase feature to process information from external sources.

- Azure SQL Database can protect sensitive information by masking certain columns to make their content readable only for authorized users.

- To avoid unauthorized reads of stored data, all Azure storage types implement encryption, which can be customized for legal requirements but cannot be avoided.

- Transmitted data is encrypted using TLS-level protocols.

- All deployments can be performed using ARM templates.

- ARM templates can be deployed in several ways: using Azure Portal or different client APIs or scripting procedures.

Summary exercise

In this exercise, test your knowledge about topics covered in this chapter. You can find answers to the questions in this exercise in the next section.

As mentioned much of this book was written during the global pandemic. Disease information becomes especially critical during times like these. Having up-to-date information is a must to better manage resources and protocols for health protection worldwide.

For this exercise, suppose that various international organizations have decided to centralize all information about diseases around the entire globe to apply artificial intelligence (AI) for evaluations and predictions. Drawing from previous experience managing global disease data, these organizations have established the following requirements:

- Data must be registered by medic and health professionals right from the source, without any intermediaries.

- All the professionals from around the globe will be invited to participate and record the required information.

- Each case must be identified by disease, symptoms, and treatments to have a follow-up.

- All attributes must be codified and organized to facilitate the entry of data with few steps — for example, by using the International Classification of Diseases, which can be extracted with its API. It is not updated very frequently, but contains hundreds of thousands of entries, with at least four levels of classification.

- Drug information must be normalized. Different countries have different legal requirements, the same medication could have different names or concentrations, and it is important to evaluate the most responsive treatments.

- In countries where technology permits, data for biological counters will be obtained by medical IoT devices and stored appropriately.

- Other statistics about social situation, population, and so on must be provided to different organizations and governments to complete the big picture, ideally in JSON or CSV format.

- Other data sources must be collected and stored to evaluate the quality of the data against several sources.

- Even when no personal identification is registered, all data-protection laws must be followed.

Now answer the following questions:

1. Which is the best storage approach for the individual entries?

   A. Azure Blob storage with GPRS replication

   B. Azure SQL Database with geo-replication

   C. Cosmos DB using Cassandra API

   D. Cosmos DB using SQL API

2. Which is the best storage approach for the International Classification of Diseases, which will be updated by the direction committee when it changes?

   A. Azure Blob storage with GPRS replication

   B. Azure SQL Database with geo-replication

   C. Cosmos DB using Table storage

   D. Cosmos DB with Gremlin API

3. Where would you store statistical information from governments and other organizations?

   A. Azure Blob storage with GPRS replication

   B. Azure Blob storage with LRS replication

   C. Cosmos DB using Table Storage API

   D. Azure Synapse Analytics pool

4. For data from IoT devices, which is your best storage option?

   A. Cosmos DB using the Table Storage API

   B. Azure Data Lake Storage (ADLS) Gen2

   C. Azure Blob storage with LRS replication

   D. Azure Blob storage with anonymous access

5. How would you manage the access of health professionals around the world to stored data?

   A. By creating a user for each person in Azure AD and adding each of them to the Contributor RBAC

   B. By creating users at the storage level and assigning each of them the Contributor RBAC

   C. By using a storage or access key from a public-facing web application with anonymous access

   D. By using a storage or access key from a public-facing web application with access management by Azure AD and external authorization like Windows Live enabled

6. How would you manage access for data scientists who will evaluate data from Azure SQL Database and propose ML procedures to predict results?

   A. By creating a user for each person in Azure AD and adding each of them to the Contributor RBAC

   B. By creating users at the storage level and assigning each of them the Contributor RBAC

   C. By configuring AD authentication in the database for an AD group assigned to the db_reader and db_denywriter database roles

   D. By using a storage or access key from a public-facing web application with access management by Azure AD and external authorization like Windows Live enabled

Summary exercise answers

This section contains the answers to the summary exercise questions in the previous section and explains why each answer is correct.

1. **D: Cosmos DB using SQL API**   You need to continuously store from different locations. Cosmos DB, with multiple writes geo-replication, allows you to make the writes quickly and return to the application.

2. **A: Azure Blob storage with GPRS replication**   The data is not a good candidate for Table storage. It contains several integration levels, and it would be difficult to query it quickly enough from a responsive application. Having data in a Blob storage with geo-replication makes it easy to upload the data to a central writing region and leaves the GPRS process to replicate the data and allow read access from the rest of the regions.

3. **B: Azure Blob storage with LRS replication**   Data will be required only during analytical processes, so there is no need to replicate it in several regions. Having a storage with LRS will be enough to protect the data, considering it could be easily obtained from the original sources in case of a data-center disaster.

4. **A: Cosmos DB using the Table Storage API**   Data from IoT devices will come in fast and in huge volumes and must be stored very quickly. Cosmos DB manages data in a few milliseconds, and you can identify each entry by device ID and timestamp.

5. **D: By using a storage or access key from a public-facing web application with access management by Azure AD and external authorization like Windows Live enabled**   Obviously, you cannot leave the entry points available for anonymous access. But it could be hard to administer millions of users in a centralized way. If you have the application authenticate users through external providers, you can receive registrations with a specific ID, and then ask a local manager to validate it. This will enable you to ensure the registering user has the professional ability to enter data before allowing Contributor-level access.

6. **C: By configuring AD authentication in the database for an AD group assigned to the db_reader and db_denywriter database roles**   This is the best way to accomplish this because you will not be in charge of maintaining the users. The organization IT department will manage which AD users belong to which groups and read rights will be assigned to users by inheritance from the AD group to the database roles.

## Chapter 3

Manage and develop data processing for Azure Data Solutions

The Azure ecosystem offers many options to process data at scale. Services like Azure Data Factory, Azure Databricks, Azure Synapse Analytics, and Azure Stream Analytics are very versatile, and are typically found in a modern data warehouse architecture.

The most common types of workloads in modern data warehouse architectures are batch processing and streaming. While both gather data from one or more sources, process it, and load it into a destination sink, they have different traits and require specific approaches.

Batch processing usually occurs on a schedule, typically at night to avoid affecting operations on source systems. Data stores that are fully operational 24/7 might require you to offload source data to an external store (for example, Azure Blob storage) to decouple the batch process from its source.

Streaming data comes in a continuous flow of events (from a sensor or a social network, for example) and is usually analyzed in specific time windows. Aggregations are performed over events that fall within the boundaries of the currently analyzed window. The result of such aggregations is displayed on a real-time dashboard for monitoring purposes or saved in a data store for further analysis. In addition, raw, non-aggregated events can be offloaded to a data store. This gives you a large amount of data to feed machine learning (ML) models as well as perform time-series and predictive maintenance analysis or anomaly detection, for example.

In a modern data warehouse, streaming and batch workloads can co-exist and work in parallel on different layers (see Figure 3-1):

- **Speed** This layer ingests streaming events, enriches them with static data if needed (for example, adding information about the devices that generated the events), aggregates them, and displays/stores the results on a dashboard or in a database.

- **Batch** This layer takes all streaming events (aggregated or not) ingested during the day, performs some transformations on them or uses them to train an ML model with fresh data, and loads them into a data warehouse.

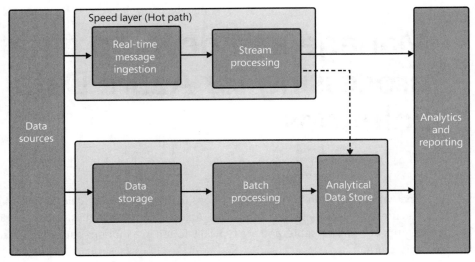

FIGURE 3-1 Speed layer versus batch layer

This is just a sample architecture, but it conveys the idea of mixed workloads in modern data warehousing. This chapter explores such workloads and related services in detail.

Topics covered in this chapter:

- Batch data processing
- Streaming data

Batch data processing

Batch workloads are very common in modern data warehouse scenarios. These are usually of two types:

- Extract-transform-load (ETL)
- Extract-load-transform (ELT)

In ETL, shown in Figure 3-2, data is extracted from sources, transformed, and loaded into the destination sink(s).

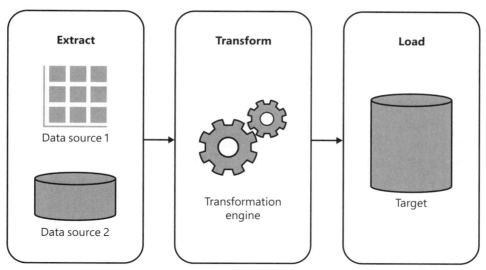

FIGURE 3-2   ETL workflow

In ELT, shown in Figure 3-3, data is extracted from sources and loaded into the destination sink. The transformation phase is performed directly at the destination. For this reason, the sink must have capabilities that enable it to work on data at scale — for example, the massively parallel processing (MPP) architecture of Azure Synapse Analytics.

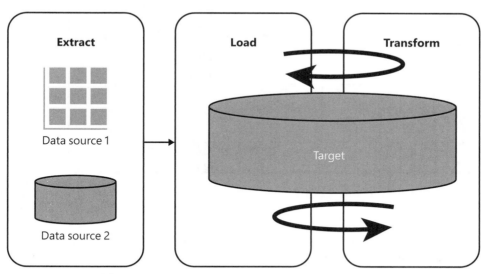

FIGURE 3-3   ELT workflow

Where streaming is dynamic, batch processing moves and transforms data at rest. If you have ever implemented a BI system, in all likelihood you have dealt with a batch workload at least once.

Figure 3-4 shows a typical batch workflow in Azure. As you can see:

1. Source data is ingested into the data store of choice, like Azure Blob storage, Azure Data Lake Storage (ADLS), Azure SQL Database, or Azure Cosmos DB.

2. Data is processed by a batch-capable engine, like Azure Data Lake Analytics, Azure HDInsight, or Azure Databricks, using a language like U-SQL, Hive, Pig, or Spark.

3. Data is stored in an analytical data store, like Azure Synapse Analytics, Spark SQL, HBase, or Hive, to serve business reporting.

NOTE Azure Data Factory or Oozie on Azure HDInsight can be used to orchestrate the whole process.

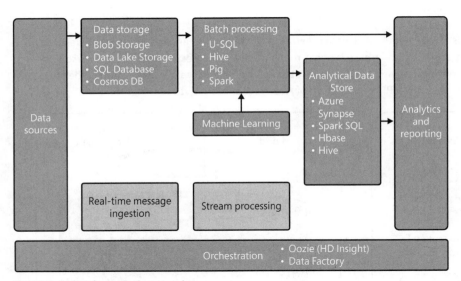

FIGURE 3-4 Batch processing overview

NOTE Azure Synapse Analytics and Azure Databricks aim to be unified platforms for ingestion, processing, storing, and serving batch and streaming workloads.

A *batch job* is a pipeline of one or more batches. These batches could be serial, parallel, or a mix of both, with complex precedence constraints. Batch jobs are usually triggered by some recurring schedule or in response to a particular event, like a new file being placed in a monitored folder. However, they mostly run off-peak to avoid incurring in-resource contention with production systems. The volume of data to be processed can be huge in scenarios falling under the (often abused) term *big data*.

This section focuses on three services that are often used in combination when implementing a batch workload in Azure: Azure Data Factory, Azure Databricks, and Azure Synapse Analytics.

Develop batch-processing solutions using Azure Data Factory and Azure Databricks

Before describing the synergy that exists between Azure Data Factory and Azure Databricks, it is worth explaining what they are (and what they are not). It is very important to understand the pros and cons of every service you implement in your cloud solution, since most of them have some areas of overlap.

For example, both Azure Databricks and Azure Data Factory have scheduling capabilities. However, while Azure Databricks allows for scheduling just a single Spark job (be it a JAR package, a Spark command, or a single notebook), Azure Data Factory allows for scheduling a whole pipeline, which can be a complex graph of interconnected activities (that can also include a Spark job performed on Azure Databricks).

The multi-purpose nature of Azure Data Factory makes it a good candidate for the orchestration of complex workloads, so you can centralize all the logging and monitoring of your activities in one place. On the other end, the scheduling capability of Azure Databricks represents a useful tool for those scenarios where this service is central to your development and deployment life cycle and you do not want to introduce components outside its scope.

Azure Data Factory

Azure Data Factory (ADF) is a cloud-based ETL and data-integration service. It allows you to create data-driven workflows for orchestrating data movement and transforming data at scale.

ADF has many strong points that elevate it to an enterprise-ready, on-demand service. To list a few:

- Hybrid scenario support, to seamlessly connect your on-premises architecture to the cloud
- Best-in-class integration with other popular services (Azure storage, Azure SQL Database, Azure Synapse Analytics, Azure Databricks, to name a few) to quickly connect all pieces together
- Visual authoring, to speed up development and maintenance
- Extensibility, to cover every possible scenario with custom connectors
- Continuous integration (CI) and continuous delivery (CD) native integration, to integrate ADF workflows with your existing DevOps pipelines

- An API layer, to control and manage it from your existing application and script suites
- A broad monitoring and alert system, to take timely action when something goes wrong

Before going into more depth on the service component, it is important to understand what ADF is not. Azure Data Factory is not a transformation engine in itself; rather, it orchestrates external services to perform data processing. ADF has only limited data-conversion capabilities when performing data movement from a source to a destination, like changing file formats or flattening out JSON nested structures. Even when you author a visual workflow using mapping data flows, ADF behind the scenes will leverage an on-demand Spark cluster to run it.

ADF can be used in every phase of an ETL/ELT process. In fact, its native integration with the most important platforms and services and its scalable data-movement engine allow for a broad range of uses, like data ingestion (the E or extract phase) and orchestration of complex workflows.

> **NOTE** Another option for orchestration is Oozie on HDInsight, but it supports only a subset of job types. Follow this link to learn more about it: *https://docs.microsoft.com/en-us/azure/hdinsight/hdinsight-use-oozie-linux-mac*.

An Azure subscription can contain one or more Azure Data Factories (think of them as ADF *instances*). Besides the obvious reasons to isolate one project from another, you may need to provision multiple instances to, for example, support multiple stages, like development, test, user acceptance testing (UAT), and production.

> **NOTE** Using the Azure Portal is not the only way to create an Azure Data Factory. Other common methods include using PowerShell, .NET, Python, REST, and ARM templates. Read more here: *https://docs.microsoft.com/it-it/azure/data-factory/introduction#next-steps*.

CREATE AN AZURE DATA FACTORY

Creating an Azure Data Factory is very easy. Follow these steps:

1. Open *https://portal.azure.com* and use your credentials to log in.
2. In the search box, type **Data Factory**, and select the corresponding entry that appears in the drop-down list.
3. In the Data Factory page, shown in Figure 3-5, click the **Create** button.
4. In the **Create Data Factory** page, in the **Basics** tab, configure the following settings:
   - **Subscription**   Choose the subscription that will contain your ADF instance.
   - **Resource Group**   Choose the resource group for your ADF instance.

FIGURE 3-5 The Data Factory page

NOTE If the resource group does not exist yet, you can create it from here.

- **Region**   Choose the desired region for your ADF instance.
- **Name**   Type a name for the ADF instance. This must be globally unique.
- **Version**   V1 is considered a legacy version and should not be used for new deployments.

5. Click the **Git Configuration** tab and configure the following settings:

- **Repository Type**   Choose **Azure DevOps** or **GitHub**.
- **Azure DevOps Account**   If you choose Azure DevOps as your repository type, enter your account info here.
- **Project Name**   Type a name for your project here. (Again, this is for Azure DevOps only.)
- **Repo Name**   Type your repository name.
- **Branch Name**   Type the branch name.
- **Root Folder**   Type the root folder path.

NOTE If you prefer, you can select the **Configure Git Later** check box if you want to set it up later.

6. Click the **Networking** tab and configure the following settings:

- **Enable Managed Virtual Network on the Default AutoResolveIntegrationRuntime**   Instruct Azure Data Factory to provision its default auto-resolve Integration Runtimes (more on that in the section "Implement the Integration Runtime for Azure Data Factory" later in this chapter) in a managed virtual network to connect to other services via a private endpoint without the burden of managing the VNET yourself. (This feature is in public preview at the time of this writing.)

- **Connectivity Method**   Choose whether to connect even your self-hosted Integration Runtime (again, more on that in the "Implement the Integration Runtime for Azure Data Factory" section) to Azure Data Factory via public endpoint or private endpoint.

7. Click the **Tags** tab and enter any custom tags you like.

> **NOTE**   Remember, tags are name/value pairs assigned to a particular resource mostly for billing consolidation.

8. Click the **Review + Create** tab. Then click the **Create** button.

The Azure Portal generates an ARM template for the Azure Data Factory, sends it to be deployed, and displays a page with a *Your deployment is in progress* message, informing you that it is currently in the creation phase.

> **IMPORTANT**   A single Data Factory is bound to the source control system as a whole. You cannot push just part of the current changes to the repository (technically speaking, you cannot cherry-pick your modifications). All your code changes must be committed together.

AUTHOR IN AZURE DATA FACTORY

After you create your Azure Data Factory instance, you can start authoring in it. To start, click the **Author & Monitor** button in the new resource's Overview page, as shown in Figure 3-6. A new browser page opens, pointing to the URL *https://adf.azure.com/home* followed by a parameter that contains the resource URI of the Data Factory you are going to author.

After signing in (you do not have to re-enter your credentials, since Single Sign-on carries them over for you), you access the multi-tenant web application that lets you develop, manage, and monitor resources and pipelines of your Data Factory. The home page of that application, which we refer to in this chapter as the Azure Data Factory authoring portal, shown in Figure 3-7, presents a quick collapsible toolbar on the left, with four menu items:

- **Data Factory**   This is the home page you are in.
- **Author**   This is where you create your pipelines.

FIGURE 3-6 Azure Data Factory Overview page

- **Monitor** This is where you monitor and analyze all executions of your pipelines or triggers.

- **Manage** This is where you can configure properties or resources that affect the whole Data Factory rather than a single pipeline, like connections to data stores, source control integration, and so on.

FIGURE 3-7 Azure Data Factory visual authoring tool home page

The central pane contains shortcuts to common tasks in Data Factory development:

- **Create Pipeline** This opens the author page with an empty pipeline ready to be edited.

- **Create Data Flow**   This opens the author page with an empty mapping data flow ready to be edited.
- **Create Pipeline from Template**   This opens the Template gallery, where you can choose between ready-to-use templates that cover many common patterns in data-pipeline development.
- **Copy Data**   This opens the Copy Data wizard, which guides you in creating a data-movement pipeline through a few simple steps (more on this in the section "Create pipelines, activities, linked services, and datasets" later in this chapter).
- **Configure SSIS Integration**   This opens the Azure-SSIS Integration Runtime Creation tab (more on this in the "Implement the Integration Runtime for Azure Data Factory" section).
- **Set Up Code Repository**   This opens the Source Control Binding Configuration tab.

> **NOTE**   Scrolling down a bit you will find a very useful feed of videos and a collection of quick start tutorials.

> **NEED MORE REVIEW?**   Templates are a convenient way to implement well-known patterns without reinventing the wheel. You can read more about them here: *https://docs.microsoft.com/en-us/azure/data-factory/solution-templates-introduction*.

DATA FACTORY COMPONENTS

Important Data Factory components include linked services, datasets, activities, and pipelines. Figure 3-8 shows the relationships between them and how they work together.

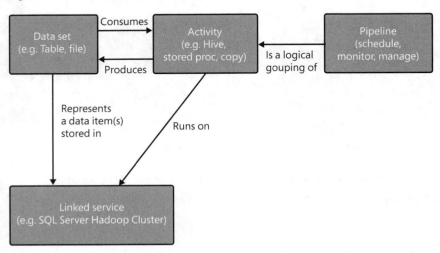

FIGURE 3-8   The logical relation between linked services, datasets, activities, and pipelines

The components in Figure 3-8 can be briefly described as follows:

- **Linked service**   A connection to a data store or service. Activities use this to actually perform the work, like copying data between stores or executing a particular job.

- **Dataset**   Represents data stored (or to be stored) on a linked service, along with its format and/or schema, if known. If the store is a database, it usually maps to a table or a view; if the store is an object store, like Blob storage or data lake storage, it is some kind of file format, like CSV, Parquet, Avro, JSON, binary, and so on.

- **Activity**   A task inside a pipeline. Activities are responsible for performing the actual work. Datasets are used to indicate the source and the sink of the activity. Depending on the type of the activity, you could have both a source and a sink (for example, with the copy activity), just one of them (for example, with the lookup activity), or none of them (for example, with the execute SSIS package activity).

- **Pipeline**   A logical grouping of activities. A pipeline represents the entry point of a Data Factory job; in fact, you cannot have activities outside a pipeline. A pipeline can be invoked manually or programmatically, or activated by a trigger. Pipelines can be nested using the execute pipeline activity, so you can physically separate the stages of your workload and orchestrate them leveraging a parent/child pattern.

> **NEED MORE REVIEW?**   To learn more about the different ways to execute a pipeline and all the available types of triggers, follow this link: *https://docs.microsoft.com/ en-us/azure/data-factory/concepts-pipeline-execution-triggers*.

Later, this chapter contains dedicated sections for all these components, in which you will learn how they work and how to create and use them.

Azure Databricks

One of the biggest limitations of the Hadoop MapReduce distributed computing paradigm is that input, output, and intermediate results are stored on disk. As data processing grows, this can quickly create a bottleneck. So, in 2011, a group of students from Berkeley, CA developed an open-source project called Spark, which works primarily in-memory. This approach was well-received, and Spark was widely adopted.

In 2013, the creators of Spark founded Databricks, a company that aims to take the engine to the enterprise-level, concentrating the whole Spark ecosystem on one single platform-as-a-service (PaaS) offering.

Databricks is a unified analytics platform. That is, it enables data engineers, data scientists, ML engineers, and data analysts to seamlessly work together in the same workspace.

> **NEED MORE REVIEW?**   To learn more about Databricks, see *https://databricks.com/*.

The Microsoft Azure team and Databricks have worked hard to create the best possible integration between the two platforms: Azure Databricks. Azure users can quickly provision an Azure Databricks service and benefit from its integrated enterprise-level security and optimized connectors to the most popular Azure components, like Azure Blob storage, ADLS, Cosmos DB, Azure Synapse Analytics, Azure Event Hub, Power BI, and more.

The main characteristics of Azure Databricks are as follows:

- **Notebooks**   You can author code using so-called *notebooks*, which may be familiar to you if you have ever used Jupyter Notebooks or the IPython interface. A notebook is a web-based interface organized in cells that enables you to assemble code, rich text (in the Markdown language, which is the same language used by the Microsoft Docs platform), and small dashboards in the same document. Notebooks allow for multi-user interactive programming, support versioning, and they can be operationalized with job-based executions.

- **Cluster Manager**   You can easily create, manage, and monitor clusters through this handy UI. Multiple runtime versions are supported, so you can create a new cluster with the latest (or beta) runtime released without affecting the existing workload. In addition, the serverless option lets you focus just on managing costs, demanding from the platform the provisioning and scaling of the resource needed to run the requested workload.

- **Optimized runtime**   Although Spark is available in Azure HDInsight as one of the cluster types, Databricks is based on a closed-source, heavily optimized version of the open-source runtime. Nonetheless, Databricks is still one of the most proficient contributors to the open-source version of Spark, and many new features and improvements are first developed internally and then released to the public. On the other hand, Spark is the only type of cluster available in Databricks.

- **Enterprise-level security**   Azure Databricks is deeply integrated with Azure Active Directory (AD), allowing for a clear separation of users and groups management (which can be conducted entirely by the IT department) and internal platform roles and authorization. For that matter, it offers very fine-grained control over what a user can do when that user works inside the environment. For example, you can restrict cluster access to specific AD groups, or you can enable users to monitor notebooks and job executions but prevent them from actually running any workloads.

> **NOTE**   Role-based access security is a Premium-only feature.

- **Delta Lake**   First developed internally by Databricks and then released to the public, Delta Lake is an open-source storage layer that runs on top of a data lake. It provides ACID transaction support and scalable metadata handling and can treat a batch table as a streaming source and sink, thereby unifying batch and streaming workloads.

- **MLOps**   Azure Databricks has full support for the whole ML life cycle. You can develop, test, train, score, and monitor your model leveraging ML Runtime (a runtime dedicated to ML that you can choose for your cluster), Spark Mlib library, MLFlow integration, and the Model Registry.

- **REST API**   You can interact with your workspace through a broad set of APIs, which makes it very easy to integrate with existing applications and workflow.

> **NEED MORE REVIEW?**   To explore all the features offered by Azure Databricks and learn more about this service, see *https://docs.microsoft.com/it-it/azure/databricks/scenarios/what-is-azure-databricks*.

> **NEED MORE REVIEW?**   Delta Lake is growing rapidly in popularity. In modern data warehousing, data lakes are very common, and the whole process benefits from making them more robust and reliable. Read more here: *https://docs.microsoft.com/en-us/azure/databricks/delta/delta-intro*.

As mentioned, Azure Databricks is based on Spark. The architectural foundation of Spark relies on the Resilient Distributed Dataset (RDD), which is:

- **Resilient**   An RDD is immutable by nature, which means that its underlying structure makes it always possible to reconstruct it in case of failure of one of the nodes of the cluster.

- **Distributed**   An RDD is a collection of objects partitioned across the nodes of the cluster, which makes it possible to parallelize and scale most of the work.

- **Dataset**   An RDD maps data stored on a filesystem (usually a distributed one like HDFS) or database. You see them like a tabular representation of the underlying data, but with cells that can contain complex objects as their values (for example, JSON nested elements). They do not contain data; they are just a pointer to it.

> **NOTE**   An RDD layer has been abstracted in later versions of Spark with the introduction of the DataFrame API in 2013 and the Dataset API in 2015. From Spark 2.*x* onward, usage of these APIs is encouraged, because they allow for more robust and clean code.

The Spark ecosystem is very wide (see Figure 3-9), and all its components are included in Azure Databricks. They are as follows:

- **Spark Core API**   Spark is built on top of Scala, but it supports five different programming languages: R, SQL, Python, Scala, and Java. The first three of these could be seen as a dialect of Spark; in fact, we talk respectively of SparkR, SparkSQL, and PySpark. In notebooks, you can use and mix all these languages, with the exception of Java. To use Java, you have to create a JAR library in your preferred IDE and import it to your workspace so you can reference in notebooks the classes defined therein.

- **Spark SQL and data frames** Spark SQL is the module for working with structured data. Your data is usually read, manipulated, and written through objects called data frames, which as we said already, are an abstraction layer over RDDs. While datasets provide a type-safe approach to the underlying data, data frames are often preferred for their higher versatility in handling dirty values, missing or unattended columns, and more.

- **Streaming** Spark provides stream-processing capabilities and integrates well with stream-ingestion engines like Azure Event Hub and Kafka.

- **MLlib** Spark comes with an ML library with common algorithms and utilities built-in, for both model development and deployment to production.

- **GraphX** This module provides support for graphs and graph computation.

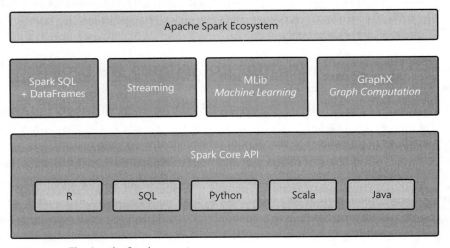

FIGURE 3-9 The Apache Spark ecosystem

PROVISION AN AZURE DATABRICKS INSTANCE

Provisioning an Azure Databricks instance from the Azure Portal requires just a few simple steps:

1. Open *https://portal.azure.com* and use your credentials to log in.

2. In the search box, type **Azure Databricks**, and select the corresponding entry that appears in the drop-down list.

3. In the Azure Databricks page, shown in Figure 3-10, click the **Create** button.

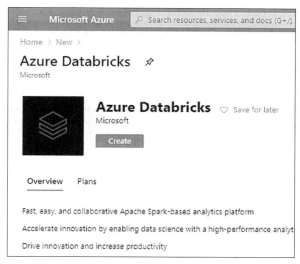

FIGURE 3-10   The Azure Databricks page

4. In the **Create Azure Databricks** page, in the **Basics** tab, configure the following settings:

    ▪ **Subscription**   Choose the subscription that will contain your Azure Databricks instance.

    ▪ **Resource Group**   Choose the resource group for your Azure Databricks instance.

    > *NOTE*   **If the resource group does not exist yet, you can create it from here.**

    ▪ **Region**   Choose the desired region for your Azure Databricks instance.

    ▪ **Name**   Type a name for the Azure Databricks instance. This must be globally unique.

    ▪ **Pricing Tier**   Choose Standard, Premium, or Trial. (You will learn more about these pricing tiers in the section "Navigate your Azure Databricks instance," in the sidebar "Azure Databricks cost considerations.")

5. Click the **Networking** tab and, under **Deploy Azure Databricks Workspace in Your Own Virtual Network (VNet)**, choose **Yes**.

    This will deploy all Azure Databricks resources in an Azure-managed virtual network (VNET), which is the default. This creates a locked resource group that contains all the needed components and services, like the managed VNET, a storage account, the nodes of the cluster, and more.

6. Specify a private VNET where you want the resources to be provisioned.

7. Click the **Tags** tab and enter any custom tags you like.

8. Click the **Review + Create** tab. Then click the **Create** button.

9. The Azure Portal generates the necessary ARM template, sends it to be deployed, and displays a page with a *Your deployment is in progress* message, informing you that it is currently in the creation phase.

> **NEED MORE REVIEW?** If you want to know more about provisioning Azure Databricks in a private VNET, see *https://docs.microsoft.com/en-us/azure/databricks/administration-guide/cloud-configurations/azure/vnet-inject*.

NAVIGATE YOUR AZURE DATABRICKS INSTANCE

After your Azure Databricks instance is provisioned, you can start navigating it. To do so, click the **Launch Workspace** button in the new resource's Overview page, shown in Figure 3-11.

FIGURE 3-11 Azure Databricks Overview page

A new browser page opens, pointing to a URL like the following: *https://adb-<uniqueid>.<#>.azuredatabricks.net/o=<uniqueid>*. After you sign in (you do not have to re-enter your credentials because Single Sign-on carries them over for you), you can access a multi-tenant web application that lets you interact with your instance. This is called the *control plane* and is hosted on the global Azure infrastructure.

Communication and interaction with the resource of your workspace is provided through *VNET peering*. This is a secure channel between two VNETs. For Azure Databricks, this secure channel is usually created between the VNET where you deployed your workspace and the VNET that hosts the control plane web application. The home page of that application is shown in Figure 3-12.

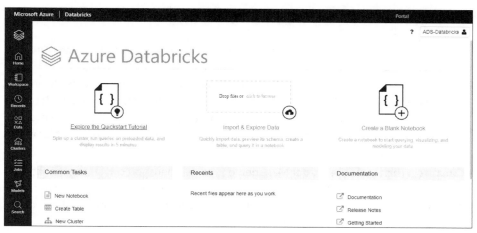

FIGURE 3-12 The Azure Databricks control plane

This home page contains a toolbar on the left, with nine buttons:

- **Azure Databricks**   This simply leads back to the Azure Databricks Overview page.

- **Home**   This opens a Workspace tab (see Figure 3-13) that contains a directory-tree view of the files and folders stored in your workspace (usually notebooks), pointing you to the defined home directory. The home directory is by default a sub-folder with the same name as the user who is logged in (typically the Azure username). It is visible only to administrators and to the owner.

> **NOTE** Administrators can restrict access to folders outside the home directory to specific users and groups.

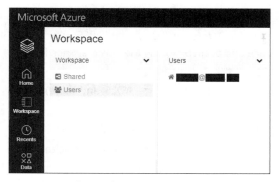

FIGURE 3-13 The Workspace tab

- **Workspace**   This is a toggle for the Workspace tab. This tab can also be pinned to keep it open.

- **Recents**   This opens a list of the most recently accessed items — for example, note-books you edited or ran.

- **Data**   Click this to create, manage, and delete databases and tables. Databases are col-lections of table objects and tables are collections of structured data. You interact with them through the Spark API—specifically with data frames. These are accessible only when there is at least one cluster running. Tables can be either global or local. Global tables are accessible from all clusters and are registered in the Hive metastore. Local tables are visible only from the cluster where they have been created. Local tables are also called *temporary views*.

- **Clusters**   This opens the Cluster Management page, where you can create, manage, monitor, and delete your clusters. Typical management tasks include changing the num-ber and the size of nodes, tuning autoscaling, setting a grace period before the nodes should auto-shutdown when not used, and so on.

> **NOTE**   See the section "Implement Azure Databricks clusters, notebooks, jobs, and autoscaling" later in this chapter for more information about clusters.

- **Jobs**   Click this to create and schedule jobs. A job consists of a single task, such as the execution of a notebook, a JAR library, or a `spark-submit` command.

- **Models**   This opens the machine learning model registry, where you can manage and track the life cycle of your models.

- **Search**   Click this to search your workspace for a specific notebook or term.

> **NEED MORE REVIEW?**   For more about credential pass-through, see *https://docs.microsoft.com/en-us/azure/databricks/security/credential-passthrough/adls-passthrough*.

Access to the storage layer is granted through the *Databricks File System* (*DBFS*), which is a distributed file system mounted into a workspace and available on all the clusters. Specifically, it is an abstraction layer, and it has the following perks:

- It allows for mounting external storage (like Azure Blob storage, ADLS, and more), so you can access it without entering credentials every time.

- You can interact with object storage with typical file semantics rather than URLs.

- You can persist files to object storage to avoid losing data after a cluster has been terminated.

Every workspace comes with a local storage (an Azure Blob storage account deployed on the managed resource groups) that can be accessed through the default storage location, the DBFS root. Figure 3-14 shows the content of this folder, obtained issuing a `%fs ls` command to a running cluster in a notebook cell. (A size value is returned for files only; folders always have a size value of 0.)

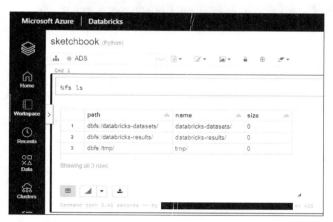

FIGURE 3-14 The content of the DBFS root folder

A special folder, /mnt, can be used to access any external storage that has been mounted on the cluster. To mount an external storage, you use the `dbutils.fs.mount` command in a notebook cell, passing authorization information in the `extra_config` parameter.

For example, after you properly set up authentication, you can list the content of a folder on an ADLS Gen2 with syntax like this:

```
dbutils.fs.ls("abfss://<file-system-name>@<storage-account-name>.dfs.core.windows.
net/<directory-name>");
```

In the same way, you can access a folder on an Azure Blob storage with syntax like the following:

```
dbutils.fs.ls("wasbs://<container-name>@<storage-account-name>.blob.core.windows.
net/<directory-name>");
```

If you want to know more about all the available approaches, follow these two links:

- *https://docs.microsoft.com/en-us/azure/databricks/data/data-sources/azure/azure-datalake-gen2*
- *https://docs.microsoft.com/en-us/azure/databricks/data/data-sources/azure/azure-storage*

For example, to mount an Azure Blob storage container named globaldata from the account ADSaccount, you can use either an access key (see Figure 3-15) or a SAS token. Mounts are accessible by any user connected to the cluster.

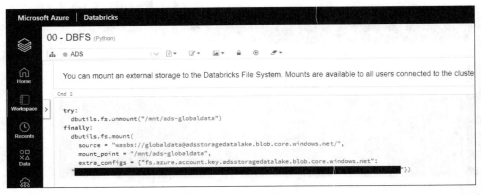

FIGURE 3-15 How to mount an Azure Blob storage container

> **NEED MORE REVIEW?** To learn more about DBFS and explore all the options available for mounts, see *https://docs.microsoft.com/en-us/azure/databricks/data/databricks-file-system*.

Azure Databricks cost considerations

Azure Databricks' billing is based on Databricks Units (DBUs), which track how many seconds the runtime actually works. In addition, you must consider the infrastructure cost for the nodes of your clusters (which are normal Azure VMs), the public IP address, and the internal Blob storage account.

Two factors affect DBU cost:

- **Tier (or SKU)** DBUs cost more on the Premium tier than on the Standard tier. The Free tier gives you zero-cost DBUs for a period of 14 days; then you have to choose between Standard and Premium.

- **Type of workload** This depends on what type of work you actually do on your clusters. Options include Data Engineering, Data Engineering Light, and Data Analytics. These names could be misleading, because they do not refer to the Spark library you use or the type of commands you issue. In fact, they generally differentiate infrastructure-type workload (for example, the actual work the internal scheduler has to do to run a job) from data workload (actually, any Spark command belongs to this category, be it a ML model training or a data frames transformation).

Discounts on DBU prices are available if you opt for a pre-purchase plan.

Azure Data Factory and Azure Databricks in a batch workload

Now that you have a high-level overview of both services, it is time to see how Azure Data Factory and Azure Databricks can co-exist and cooperate in a modern data warehouse scenario or, more in general, in a batch workload in Azure.

Typically, you can have three kinds of architecture:

- **Azure Data Factory only** In this architecture, the ingestion and transformation of source data and the orchestration of the workload are achieved through ADF. Copy activity and mapping data flows have key roles in this solution because they allow for the ingestion and transformation of the source data in a very flexible way. Both are discussed in the following sections. This is mostly a code-free approach because the whole workflow can be created visually in the ADF authoring portal.

- **Azure Databricks only** The deep integration between Azure Databricks and the other Azure services makes it possible to use Spark as an ingestion and transformation engine. For example, integration with Azure Key Vault and the support for ADLS credential pass-through enable you to secure access to your data stores. Also, optimized connectors to external services (like the one for Azure Synapse Analytics) grant best-in-class performances even with high data volume. The workflow is orchestrated through the built-in job scheduler, which usually executes a parent notebook that calls children notebooks in the required order.

- **Azure Data Factory and Azure Databricks** This is the most flexible solution among the three. Transformations of the data are performed in Spark through Azure Databricks, while ADF orchestrates the workflow and optionally extracts data from the sources (via the copy activity). In addition, the many connectors supported by ADF can be of great help when you need to integrate data from non-traditional stores — for example, REST services. However, if you do not plan (or want) to manage an Azure Databricks workspace or to write Spark code directly, the first option would be a better fit for you. In fact, mapping data flows "translate" all activities in their pipeline to custom Spark code, which is executed on an on-demand Azure Databricks cluster. On the other hand, editing Spark notebooks yourself gives you access to the full potential of Spark, and managing the clusters directly allows for finer tuning of performance.

While data movement and workload orchestration through Azure Data Factory are very common in modern data warehouse solutions, the choice of the transformation engine may be driven by some factors. For example, when opting for a Spark-based engine like Azure Databricks or mapping data flows, it is advisable to have at least a basic understanding of how Spark works. Concepts like memory occupancy, exchange partitioning, narrow and wide transformations, or actions may come into play when you need to fine-tune your input datasets for performance.

If you have a SQL background, Azure SQL Database may seem like a straightforward choice. However, if you need scale-out capabilities, it is not the most convenient option. In this case, Azure Synapse Analytics, with its MPP architecture, would fit perfectly in your architecture. But in a similar way to Spark, MPP architecture requires a specific data organization to perform well. You should have a good grasp of concepts like table distribution, the Data Movement Service (DMS), and columnar indexes when choosing this service.

The following sections go more into detail on the key components of Azure Data Factory and Azure Databricks.

Implement the Integration Runtime for Azure Data Factory

The core of Azure Data Factory is the *Integration Runtime* (*IR*). It is the compute infrastructure used to provide integration capabilities.

You can have three different types of IR, and you can create more than one IR for each type if needed:

- **Azure IR**   This is the basic IR. There must be at least one of this kind in an Azure Data Factory instance. It is the engine that actually performs data movements between cloud data stores. It is also in charge of dispatching external activities in public networks and executing data flows. It has great elasticity, and you can control how much it can (or should) scale by tuning the Data Integration Unit (DIU) properties for each activity in your pipeline that uses it.

- **Self-hosted IR**   This can be used to solve two problems:
  - You have resources in a private network or behind a firewall that do not face the internet.
  - You have data stores that require a bring-your-own driver, such as SAP HANA, MySQL, and so on.

  This IR is usually installed on one (or more, to enable scale-out) VM inside your private network and is subsequently linked to your Azure Data Factory through the creation of an additional self-hosted IR.

- **Azure-SSIS IR**   This IR supports the execution of traditional SQL Server Integration Services (SSIS) in an on-demand cloud environment. It supports both Standard and Enterprise editions. It comes in really handy in those cases when you have an on-premises workload based on SSIS packages and you want to "lift-and-shift" it to Azure PaaS with minimal effort. After you set up an Azure-SSIS IR, you can execute the deployed SSIS packages in your ADF pipelines through the execute SSIS package activity.

> **NOTE**   DIUs are a key concept in Azure Data Factory both for tuning performance and for predicting costs. Learn more here: *https://docs.microsoft.com/it-it/azure/data-factory/ copy-activity-performance-features#data-integration-units*.

Manage and configure IRs

You can manage and configure IRs in a dedicated section of the ADF authoring portal. As a reminder, you access the authoring portal from the Overview page of your Data Factory in the Azure Portal by clicking on the Author & Monitor button. (Refer to Figure 3-6.)

The toolbar on the left side of the authoring portal contains a **Manage** button (with a blue toolbox icon). Clicking it opens the Management settings for your Data Factory. Click the **Integration Runtimes** option under **Connections** to open the page shown in Figure 3-16. This page lists all the existing IRs for this Azure Data Factory instance.

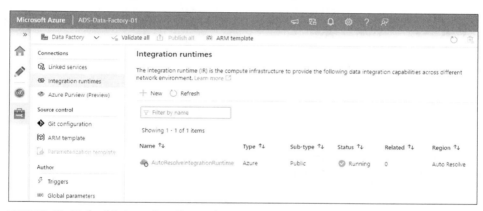

FIGURE 3-16   IRs for this Azure Data Factory instance

When you create a Data Factory, an IR named AutoResolveIntegrationRuntime is automatically set up. This IR, which cannot be removed or changed, can support all your basic workload needs. In fact, as you can see in Figure 3-16, the Region option is set to Auto Resolve, which means that when performing data movements, it automatically selects the best Azure region based on the location of the source and the sink.

For example, imagine a Data Factory located in Western Europe, and a copy activity that moves data from a storage account (source) in the Western United States to an Azure SQL Database (sink) in the same region. In Azure, all intra-region and inbound traffic moves free of charge, while extra-region outbound traffic over a certain threshold is subject to a fee. In this example, if the Azure Data Factory were to use its deployment region as a bridge between the source and sink, you would pay for both the traffic from the Western U.S. to Western Europe (first hop), and from Western Europe to the Western U.S. (second hop). However, with Auto Resolve set, the traffic would remain in the Western U.S., keeping the cost and the network lag as low as possible.

Create a new IR

Clicking the **+New** button near the top of the Integration Runtimes page opens the Integration Runtime Setup pane on the right of the screen. (See Figure 3-17.) This pane gives you two choices for creating a new IR:

- **Azure, Self-Hosted**   Choose this to create an Azure IR or self-hosted IR or to link an existing self-hosted IR in another Azure Data Factory.
- **Azure-SSIS**   Click this to create an Azure-SSIS IR.

FIGURE 3-17   The Integration Runtime Setup pane

CREATE AN AZURE IR

To create an Azure IR, follow these steps:

1. Click the **Azure, Self-Hosted** option in the **Integration Runtime Setup** pane and click **Continue**.
2. Choose **Azure IR**.
3. Type a name for your Azure IR in the **Name** box. This name must be unique within the Data Factory.
4. Optionally, type a brief description of this Azure IR (for example, its purpose) in the **Description** box.

5. Open the **Region** drop-down list and choose **Auto Resolve** to give your IR the greatest flexibility.

   You use the options in the Data Flow Runtime section to tune the cluster used for mapping data flows execution.

6. Open the **Compute Type** drop-down list and choose **General Purpose, Memory Optimized**, or **Computer Optimized**. This will be the compute type for the VMs.

7. Open the **Core count** drop-down list and select the desired option.

   This represents a combination of cores for the executors and the driver.

8. Open the **Time to Live (TTL)** drop-down list and select a value (in minutes).

9. Click the **Create** button.

TTL

The TTL is very important when you have a pipeline with two or more mapping data flows activities that run in sequence. In fact, as soon as a mapping data flows activity completes its execution, the Spark cluster used for the actual work is shut down unless the TTL has a value greater than 0. Because warming up the cluster could take some minutes, it is very ineffective to repeat its provisioning for every mapping data flows activity in your pipeline. Setting a proper value (depending on your workload) for the TTL option can ensure that the warmup happens only once within a pipeline execution. As usual for PaaS resources, it is a trade-off between execution time and cost, because you are billed for the seconds of uptime of the cluster.

CREATE A SELF-HOSTED IR

To provision a self-hosted IR, follow these steps:

1. Click the **Azure**, **Self-Hosted** option in the **Integration Runtime Setup** pane and click **Continue**.

2. Choose **Self-Hosted**.

3. Type a name for your self-hosted IR in the **Name** box. This name must be unique within the Data Factory.

4. Optionally, type a brief description of this self-hosted IR (for example, its purpose) in the **Description** box.

For an Azure IR, the background infrastructure is managed by Azure. But for a self-hosted IR, you must set up and add the node(s) composing the runtime yourself. First, you must install an MSI package on the VM(s) that will act as nodes for the IR.

5. Download the MSI package from the setup pane or from *https://www.microsoft.com/ en-us/download/details.aspx?id=39717*.

6. Register your node(s) to the self-hosted IR you just created. (The MSI package handles this for you.)

7. Specify whether the IR package will auto-update itself to newer releases or whether this will be done manually.

8. Specify whether you want to share this self-hosted IR with other Azure Data Factories.

For self-hosted IRs, there are some important considerations:

- Self-hosted IRs are typically used to access resources behind a corporate firewall that cannot face the internet directly.

- Self-hosted IRs are installed on one or more VMs in the private network and represent the only communication channel between Data Factory and your internal resources. This communication is outbound-only and goes from the private network to the Data Factory.

- It is not necessary to install the IR on the machine hosting the source data store. Instead, it is better to make IR nodes work as a gateway between Data Factory and your private data source. Multiple sources can be reached through the same nodes.

- The IR does not handle scaling for you. If you set up multiple nodes, they have to all be up and running to allow for scaling the workload. Warming up and shutting down additional nodes is totally up to you.

- A VM that hosts the IR could quickly become a bottleneck for your workload. In fact, it will perform the major work — for example, compressing the data while copying it, or converting the format. It is a best practice to regularly monitor the availability of resources like CPU, RAM, and disk space using the centralized monitoring capabilities of Azure Data Factory. (More on this in Chapter 4.)

CREATE AN AZURE-SSIS IR

To create an Azure-SSIS Integration Runtime (Azure-SSIS IR), follow these steps:

1. Click the **Azure-SSIS** option in the **Integration Runtime Setup** pane and click **Continue**.

2. Type a name for your Azure-SSIS IR in the **Name** box. This name must be unique within the Data Factory.

3. Optionally, type a brief description of this Azure-SSIS IR (for example, its purpose) in the **Description** box.

> **NOTE** The Type setting dictates the type of your IR. In this case, it contains the value Azure-SSIS, and cannot be changed.

4. Open the **Location** drop-down list and choose the Azure region that will host the nodes for the IR.

5. Use the **Node Size** setting to specify the size of the VM(s) that will comprise the IR.

6. Use the **Node Number** setting to specify the number of VM(s) that will compose the IR. This setting allows values from 1 to 10.

7. Open the **Edition/License** drop-down list and choose **Standard** or **Enterprise**.

   This will determine which features and components will be available in your SSIS packages, in the same way as in on-premises installations.

8. Set the **Save Money** option to **Yes** if you have a license that is eligible for Azure Hybrid Benefit for SQL Server.

9. Choose **Package Deployment Model** or **Project Deployment Model**.

   If you opt for the latter, you will be asked for a SQL Server to store the SSIS Catalog database and its service tier. Because there can be a lot of communication between the Integration Services engine and the SSIS Catalog database (think about logging, for example), it is highly recommended to have them located in the same Azure region.

   Whether you choose the package or the project deployment model, you can also create a package store hosted on a file store or an Azure SQL Managed Instance.

   The last step requires you to specify a few other options, like the maximum number of parallel executions allowed on a single node and the optional location of a script that will be used to customize the nodes when provisioned. For example, if you use third-party components in your SSIS packages, they have to be installed on the nodes of the IR to make your packages work as expected.

> **NOTE** If you want to migrate your SSIS packages to the Azure-SSIS IR, see *https:// docs.microsoft.com/en-us/azure/data-factory/scenario-ssis-migration-overview*.

NEED MORE REVIEW? You can read more about IRs here: *https://docs.microsoft.com/ en-us/azure/data-factory/concepts-integration-runtime.*

Create pipelines, activities, linked services, and datasets

Now that you have a basic understanding of IRs, you can start creating and editing pipelines. You create pipelines in the main section of the ADF visual authoring portal. As a reminder, you access the authoring portal from the Overview page of your Data Factory in the Azure Portal by clicking the **Author & Monitor** button. (Refer to Figure 3-6.) Then, in the left toolbar, click the **Author** button (the one with a pencil icon) to open the page shown in Figure 3-18.

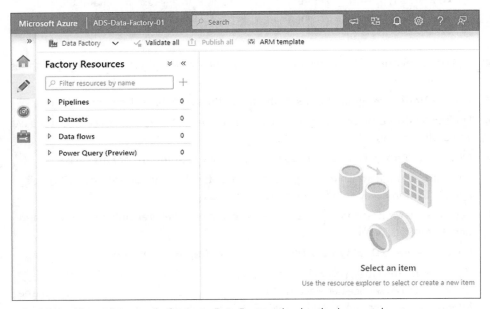

FIGURE 3-18 The author page in the Azure Data Factory visual authoring portal

Along the top of the author page are various buttons, including a Publish All button. (This button is grayed out because there is no pipeline currently being authored.) There is also a Data Flow Debug switch, which is currently off, and an ARM Template menu, which you can use to export the whole Data Factory ARM template or import an ARM template of a previously exported Data Factory. The left panel, labeled Factory Resources, is a resource explorer that you can use to create new pipelines, datasets, or data flows, and navigate or search for existing ones.

Create a pipeline

Click the **plus sign** next to the search box in the Factory Resources panel and select **Pipeline** from the menu that appears. A new tab containing a blank pipeline is added on the right. (See Figure 3-19.)

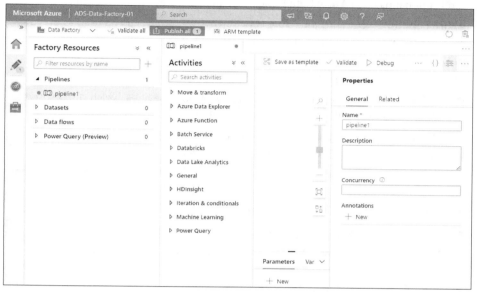

FIGURE 3-19 The pipeline creation page in the Azure Data Factory visual tool

On the left side of this tab is an Activities pane, which contains a list of activity categories. You can expand these categories and then drag and drop activities within them onto the pipeline canvas to the right. Above the pipeline canvas are buttons to save, validate, debug, and trigger the pipeline. Below it is a menu that you can use to set properties and parameters that affect the pipeline's activities. Finally, on the right is a Properties pane, where you can set the pipeline's name and description, and limit the concurrency of executions at runtime (in case multiple invocations are issued at the same time or while the pipeline is already running).

> **NOTE** Parameters in Azure Data Factory offer a simple yet powerful way to make your Data Factory instances reusable. Recently, the development team introduced global parameters, which speed up the development of nested pipelines by removing the hassle of redefining them at each level to make them bubble up to the outer level. The following whitepaper is a must-read to master parameters mechanics: *https://azure.microsoft.com/en-us/resources/azure-data-factory-passing-parameters/*.

Following is a short description of the categories of activities you can use in your pipeline:

- **Move & transform**   Contains activities for data movement and transformation, like copy data and data flow.
- **Azure Data Explorer**   Contains the Azure Data Explorer activity, which can be used to send commands to an Azure Data Explorer cluster.
- **Azure Function**   Contains the Azure Function activity, which can be used to execute an existing Azure function to run custom code.
- **Batch Service**   Contains the custom activity, which can be used to execute custom code deployed on the Azure Batch service.
- **Databricks**   It contains the Notebook, Jar, and Python activities, which can be used to issue Spark jobs to an either provisioned or on-demand Databricks cluster.
- **Data Lake Analytics**   Contains the U-SQL activity, which can be used to dispatch U-SQL jobs to the Azure Data Lake Analytics PaaS service.
- **General**   Contains several mixed-purpose activities. The most notable of these include the following:
  - **Execute pipeline**   Nests pipeline executions
  - **Execute SSIS package**   Executes Integration Services packages on an Azure-SSIS IR
  - **Stored procedure**   Executes a stored procedure contained in a cloud or an on-premises SQL Server database
  - **Web**   Calls a custom REST endpoint
- **HDInsight**   Contains the Hive, MapReduce, Pig, Spark, and streaming activities. These can be used to issue different types of Hadoop jobs to a provisioned or on-demand HDInsight cluster.
- **Iteration & conditional**   Contains the filter, for each, if condition, switch, and until activities. These can be used to control or alter the flow of the activities in your pipeline.
- **Machine Learning**   Contains the machine learning batch execution, machine learning update resource, and machine learning execute pipeline activities, which can be used to interact with the Azure Machine Learning Studio (classic) service.
- **Power Query**   Contains the Power Query mash-up activity, which can be used to perform data wrangling at scale in a Power BI Desktop–like interface.

> *NEED MORE REVIEW?*   For a deeper look into Data Factory activities, see *https://docs. microsoft.com/en-us/azure/data-factory/concepts-pipelines-activities*.

Figure 3-20 shows how the author page changes after you drag and drop the copy data activity into the pipeline canvas and select it.

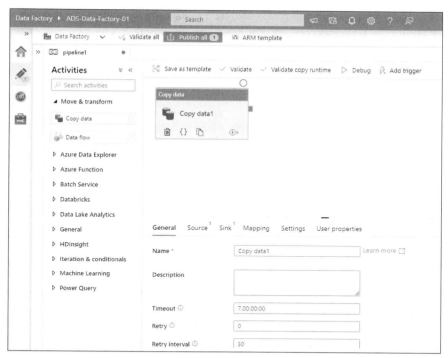

FIGURE 3-20 The copy data activity

Copy data is the main activity for performing data movement. It requires you to specify a source and a destination. The bottom pane that appears, whose options change depending on what type of item selected, reflects this, and can be used to fine-tune the activity. It contains six tabs:

- **General**   Set common properties of the activity, like its name, description, and timeout value. You can also specify whether the activity should be retried if an error occurs, the timeout between retries, and whether activity input and output should be secured (through the Secure Input and Secure Output properties). Because activity input and output are logged in plain text as JSON objects, securing them can be useful to avoid disclosing sensitive information. In fact, when secured, input and/or output are not captured at all. This tab stays pretty much the same for the majority of Data Factory activities.

- **Source**   Select the activity's source dataset from a drop-down list. If the source dataset does not yet exist, you can click the plus button to open a new tab where you can create it. (Dataset creation is covered in the next section.) For example, the tab shown in Figure 3-21 reflects a dataset that maps a SQL Server table. From here, you can preview the source data, define a timeout and isolation level for data retrieval, and enable (or not) the use of partitions to parallelize read operations. (These can be physical partitions, if present, or based on a dynamic range. Beware that this second option is prone to performance problems when there is no index that covers the target field.) You can also add extra columns based on expressions, static values, and reserved variables (like $$FILEPATH for file-based sources) and choose query settings.

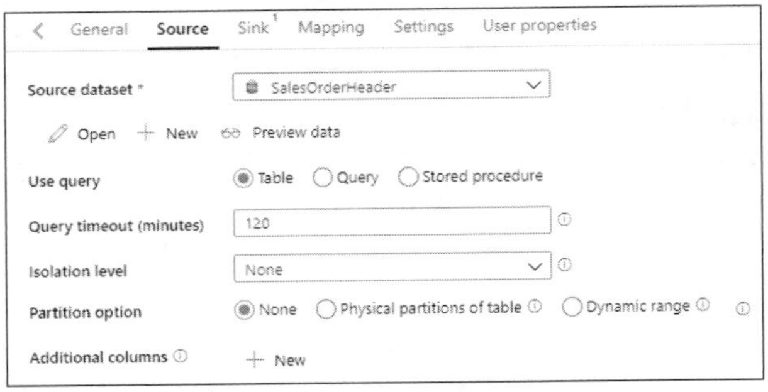

FIGURE 3-21 Dataset options for a SQL Server table

Use Query settings

The Use Query settings, specific to database sources, has interesting values from which to choose. The Table option is straightforward, but at a first glance, the Query and Stored Procedure options might not seem applicable (or seem only partially applicable) to a dataset that points to a table. The reality is, activities can use a dataset as a simple bridge to the underlying linked service — in this case, the SQL Database server — and issue commands that are unrelated to the object mapped to the dataset. For example, you can completely ignore the SalesOrder-Header table and instead query the source for the SalesOrderDetail table or call an existing stored procedure. Although this flexibility may seem like a good thing, it is always better to have a dataset with a clear scope to avoid unintentionally messing up activities later in the development process.

- **Sink** Select the destination dataset in the same way you do the source dataset.
- **Mapping** Optionally define a mapping between the source and sink, either importing them from the datasets or entering them manually. Also, optionally select a subset of the source fields. If you leave this tab empty, a schema is inferred at runtime. It is not uncommon to have loose schemas in the extract phase in modern data warehouse scenarios, since sources may be disparate and could change without notice. If you follow this approach, data integrity and schema validation are often asserted in an early stage of the subsequent data process step. A typical example is when data engineers extract data from an Enterprise data warehouse to make it available to the data science team. In this case, datasets may be very wide, because data scientists need to analyze as many fields as possible in the features-selection and engineering phases to determine whether the ML model would benefit from them. Defining and maintaining a schema of the source tables can be very time-consuming and might not provide any real advantage since fields could be ignored or transformed afterward.

- **Settings**   Tune performance-related parameters, including settings that relate to the degree of parallelism (to a maximum of 8), whether data integration units should be fixed or chosen automatically by the engine, fault tolerance settings (like whether to skip incompatible rows for database sources or skip missing files for file-based sources), and staged copy.

- **User properties**   Set name/value properties, which can be useful to better track your activity in monitoring logs.

> *NEED MORE REVIEW?*   To learn more about staged copy, including typical use cases, see *https://docs.microsoft.com/en-us/azure/data-factory/copy-activity-performance-features#staged-copy.*

Create a dataset and connectors

To create a dataset, click the **plus sign** next to the search box in the **Factory Resources** panel and select **Dataset** from the menu that appears. A tab opens on the right (see Figure 3-22), asking you to select a data store to proceed.

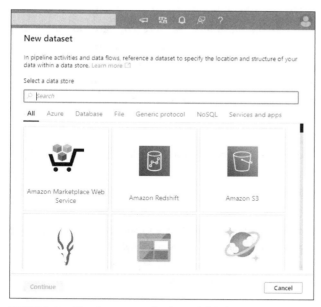

FIGURE 3-22   The data store selection tab for dataset creation

Datasets are closely related to Data Factory connectors, which enable the service to interact with remote stores in many ways. The link to the data store is the linked service component, which contains all the relevant information to connect to it, like its URL and authentication information. You can think of a dataset as a named view that points to or references data on the remote store, so activities can actually use it.

By choosing a connector, you tell Data Factory which type of dataset it has to handle. Following is a list of categories of connectors and a short description of each one.

- **All**   An unfiltered view of all available connectors.
- **Azure**   Contains all the supported Azure services, like Azure Blob storage, ADLS Gen1 and Gen2, Azure Cosmos DB (Mongo and SQL API), Azure SQL Database, Azure Managed Instance, Azure Synapse Analytics, and more.
- **Database**   Contains all the supported database services. Here, you can find both Azure and third-party services, like Amazon Redshift, IBM Db2, Google BigQuery, Hive, Oracle, Netezza, SAP Business Warehouse (BW), SAP HANA, Spark, Teradata, Vertica, and more.
- **File**   Contains file services like FTP, SFTP, Amazon S3, Google Cloud Storage, generic HTTP endpoints, and more. Also, you can reach file shares through the file system connector — both publicly available, like Azure File shares, and on-premises or private networks, through the self-hosted IR.
- **Generic protocol**   Contains more broad-use connectors, like ODBC, OData, REST, and SharePoint Online List.
- **NoSQL**   Contains connectors to NoSQL sources. At the moment, it lists Cassandra, MongoDB, and Couchbase (in preview).
- **Services and apps**   Contains connectors to popular PaaS and SaaS services, like Amazon Marketplace, Microsoft Dynamics 365, Microsoft Dynamics AX, Microsoft Dynamics CRM, Atlassian Jira, Office 365, Oracle Eloqua (in preview), Oracle Responsys (in preview), Oracle Cloud (in preview), PayPal (in preview), Salesforce, SAP ECC, Snowflake, and more.

Some connectors require you to provide additional information. The Azure Blob storage connector, for example, has to know which file format the dataset is mapping to so it can offer the proper options in the configuration pane (see Figure 3-23). Selecting the JSON format, for example, leads you to another tab where you specify the dataset name and the linked service that points to the data store. If the linked service does not yet exist, you can create it by clicking the plus sign. An integration service creation tab appears (see Figure 3-24).

Select format
Choose the format type of your data
Avro
Binary
CSV
DelimitedText
Excel
JSON
Json
ORC
Parquet
XML
XML

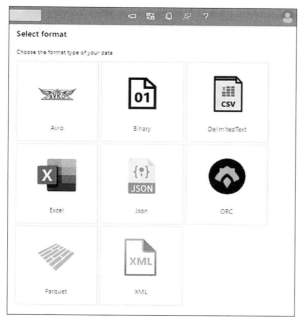

FIGURE 3-23 The file format specification selection screen for the Azure Blob storage connector

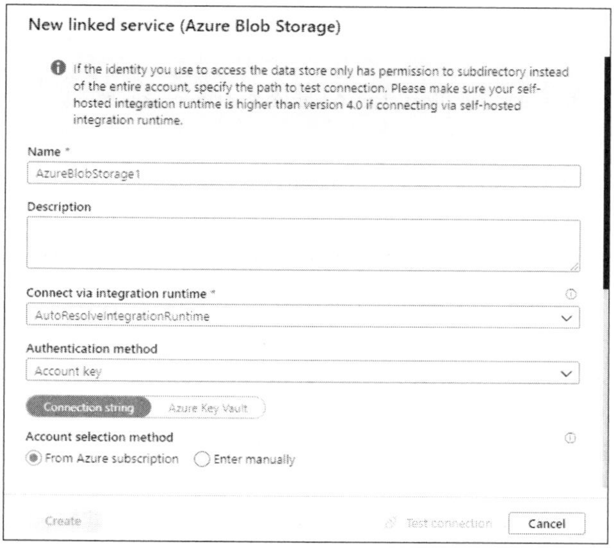

New linked service (Azure Blob Storage)
If the identity you use to access the data store only has permission to subdirectory instead of the entire account, specify the path to test connection. Please make sure your self-hosted integration runtime is higher than version 4.0 if connecting via self-hosted integration runtime.
Name *
AzureBlobStorage1
Description
Connect via integration runtime *
AutoResolveIntegrationRuntime
Authentication method
Account key
Connection string
Azure Key Vault
Account selection method
From Azure subscription
Enter manually
Create
Test connection
Cancel

FIGURE 3-24 The linked service creation screen

Notice that in the title of the tab shown in Figure 3-24, "Azure Blob Storage" appears in parentheses. This is because the type of this linked service is directly related to the kind of dataset we are creating. If you were to instead create a linked service from the Manage section of the Data Factory instance, you would see a tab with all possible connectors to choose from.

So, in addition to the Data Store section, you would see a Compute section with connectors specific to data processing (see Figure 3-25), like Azure Databricks, Azure HDInsight, and more.

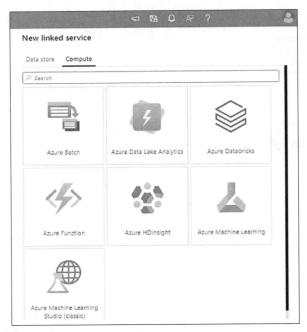

FIGURE 3-25 The available connectors for data processing

Referring back to Figure 3-24, to create an Azure Blob storage linked service, you must provide the following information:

- **Name** This must be unique within a Data Factory.
- **Description** Adding a description is optional.
- **Connect Via Integration Runtime** Here, you choose the IR used by the component. You can also create a new one if needed.
- **Authentication Method** This can be one of the following:
  - **Account Key** If you choose this, you must provide a connection string to your storage account either manually (selecting it from a subscription to which you have access) or through an Azure Key Vault secret. (See the upcoming note for more information on Azure Key Vault.)
  - **SAS URI** If you choose this, you must provide a SAS URL/SAS token pair, either manually or through Azure Key Vault secrets.
  - **Service Principal** This requires you to provide your storage account endpoint and type (either manually or by selecting it from a subscription to which you have access), your service principal tenant, ID, and key (this can be retrieved via Azure Key Vault), and the Azure cloud type it is registered to.

- **Managed Identity** This requires you to provide your storage account endpoint and type, either manually or by selecting it from a subscription to which you have access.

- **Annotations** These are custom name/value pairs to be associated to the resource.

- **Advanced** Here you can set properties not yet exposed by the UI, expressing them in JSON format.

> **NOTE** Azure Key Vault enables you to centralize the storage of sensitive information, like keys, certificates, and secrets, in a highly secure repository. Data Factory is deeply integrated with Key Vault, and pretty much all the available linked services support it. To learn how to enable and use Azure Key Vault in your Data Factory, see *https://docs.microsoft.com/en-us/azure/data-factory/store-credentials-in-key-vault*.

After you fill all the required fields, you can test the connection to the storage account and/or click the **Create** button to proceed with its creation. If no errors arise, you will be returned to the previous page to complete the definition of the JSON dataset. Notice that this page now contains two additional fields:

- **File path** Optionally specify a container, folder, and/or file for the dataset. Wildcards are accepted. Leave this empty if you want to just point to the storage account and give more freedom to the activity that will use the dataset. For example, you can specify just the container name and the root folder, and then in a copy data activity, recursively iterate its children to move the directory content in its entirety.

- **Import schema** If you specified a full file path, you can decide whether to import the file schema. If you want to import it, you can get it from the file you are pointing to or from a local sample file. This last option comes in very handy when, for example, you are creating the destination dataset of a copy data activity and the target does not yet exist. In this case, you might want to prepare a sample file with the same structure as your destination and use it to instruct the dataset about the file schema.

Click the **OK** button to create your dataset and open it in edit mode. As shown in Figure 3-26, this interface is pretty much identical to the interface for editing pipelines. It features the following tabs:

- **Connection** Contains many properties you already set in the creation process, but also introduces new ones like Compression Type and Encoding.

- **Schema** Contains the dataset schema, if already imported. You can also edit the current schema, import it again (or for the first time), and clear the existing one.

- **Parameters** Contains dataset parameters, which can be used inside expressions to customize the dataset behavior. Parameters are set by pipeline activities that use the dataset. For example, think about a dataset with a parameterized path that can be reused in different copy data activities to write the same file type in different sink locations.

FIGURE 3-26 The dataset edit window

> **NOTE** Data Factory expressions and functions are very powerful and can add great versatility to your pipelines. Read more here: *https://docs.microsoft.com/en-us/azure/data-factory/ control-flow-expression-language-functions.*

PRACTICE Use the Copy Data wizard to create and run a simple pipeline

This practice exercise shows you how to use the Copy Data wizard to create a simple yet complete pipeline from scratch. The purpose of this pipeline is to convert the JSON COVID daily data to Parquet, which is a more suitable format for Azure Databricks. Exercise 2.3 and Exercise 2.4 in Chapter 2 covered uploading files containing this data; if you want to complete this step-by-step procedure, you will need to have completed those exercises. To begin, you will provision a Data Factory, and then step through the Copy Data wizard.

> **NOTE** In the next sections of this chapter, you will use Spark to do some work on these files.

1. Open *https://portal.azure.com* and use your credentials to log in.
2. Click **+Create a Resource.**
3. Type **Data Factory** in the search box and select the corresponding entry that appears in the drop-down list.
4. Click **Create**.
5. In the **Create Data Factory** page, in the **Basics** tab, configure the following settings:
   - **Subscription** Choose the subscription that will contain your ADF instance.
   - **Resource Group** Choose the resource group for your ADF instance.

- **Region** Choose the same region you used in previous practice exercises (for example, North Europe).
- **Name** Type a name for the ADF instance. This must be globally unique.
- **Version** Leave V2 as the version.

6. Click the **Git Configuration** tab and select the **Configure Git Later** check box.

7. Skip the **Managed Identities** and **Encryption** tabs.

8. Click the **Networking** tab and configure the following settings:
   - **Enable Managed Virtual Network on the Default AutoResolveIntegrationRuntime** Leave this unchecked.
   - **Connectivity Method** Set this to **Public Endpoint**.

9. Click **Review + Create**.

10. If your new Data Factory is successfully validated, click **Create**.

11. After the Data Factory is provisioned, navigate to its **Overview** page and click the **Author & Monitor** button.

   The Data Factory UI home page opens.

12. Click the **Copy Data** option.

   The Copy Data wizard starts.

13. In the Copy Data wizard's **Properties** page, type **JSON to Parquet** in the **Task Name** box, open the **Task Cadence or Task Schedule** drop-down list and select **Run Once Now**, and click **Next**.

14. In the Source Data Store page of the Copy Data wizard, click **+Create New Connection**.

   A New Linked Service pane opens.

15. Type **Azure Blob storage** in the search box, click the **Azure Blob storage** icon that appears, and click **Continue**.

   The New Linked Service pane changes to contain several configurable settings.

16. Type **adsstoragedatalake** in the Name box.

17. Leave **Connect Via Integration Runtime** set to **AutoResolveIntegrationRuntime** and **Authentication Method** set to **Account Key**.

18. Make sure **Connection String** is selected and choose **From Azure Subscription** in the **Account Selection Method** section.

19. Select your subscription and the storage account that hosts your COVID data JSON files.

   Data Factory will try to obtain the connection string of the storage account and save it (encrypted) in the linked service.

20. Click the **Test Connection** link at the bottom of the page.

21. If everything works as expected, click the **Create** button at the bottom of the page. If not, make sure the storage account properties and authentication information are all correct.

22. Back in the **Source Data Store** page, select the **adsstoragedatalake** connection icon and click **Next**.

23. In the **Folder Path** box, enter the path where the daily JSON files reside — in this example, **globaldata/Daily/**, leave the suggested values for the other settings as is, and click **Next**.

    The File Format Settings page opens.

24. Open the **File Format** drop-down list and select **JSON Format**, leave the suggested values for the other settings as is, and click **Next**.

    The Destination Data Store page opens.

25. Select the **adsstoragedatalake** connection icon and click **Next**.

    The Choose the Output File or Folder page opens.

26. Type **staging/Daily/parquet/** in the **Folder Path** box, and click **Next**.

    The File Format Settings page opens.

27. Open the **File Format** drop-down list and select **Parquet Format**, leave the suggested values for the other settings as is, and click **Next**.

The Schema Mapping page opens.

28. Enable the **Advanced Editor** toggle button.

29. If the table does not already contain an editable row (represented by a field under the Source column), click the **+New Mapping** button to add one.

Next you will specify which fields from the source JSON to carry over to the destination Parquet file. In this practice exercise, you will map six fields in total.

30. In the editable row, type "**$['date']**" (this is a typical JSONPath syntax) in the **Source** field, type "**date**" in the **Destination** field, and select "**DateTime**" from the **Type** drop-down list.

> **NOTE** Depending on your screen resolution, you may have to scroll to the right to see all the columns.

31. Click the **+New Mapping** button to create a new editable row. Then type "**$['total_cases']**" in the **Source** field, type "**total_cases**" in the **Destination** field, and select "**Int32**" from the **Type** drop-down list. (If the "Int32" option is not visible in the list, you can scroll down the list or you can type "int" in the search text box to find it.)

32. Click the **+New Mapping** button to create a new editable row. Then type "**$['new_cases']**" in the **Source** field, type "**new_cases**" in the **Destination** field, and select "**Int32**" from the **Type** drop-down list.

33. Click the **+New Mapping** button to create a new editable row. Then type "**$['total_deaths']**" in the **Source** field, type "**total_deaths**" in the **Destination** field, and select "**Int32**" from the **Type** drop-down list.

34. Click the **+New Mapping** button to create a new editable row. Then type "**$['new_deaths']**" in the **Source** field, type "**new_deaths**" in the **Destination** field, and select "**Int32**" from the **Type** drop-down list.

35. Click the **+New Mapping** button to create a new editable row. Then type "**$['Country']**" in the **Source** field, type "**country**" in the **Destination** field, and select "**String**" as from the **Type** drop-down list.

> **NOTE** Be sure to capitalize Country in the Source field. It must match exactly the header of the source file, or it will end up as null in the output.

36. Leave the suggested values for the other settings as is and click **Next**.

The Settings page opens.

37. Leave the default settings as is and click **Next**.

The Summary page opens.

38. Verify that your settings are correct and click **Next**.

> **NOTE** The wizard allows you to change the name of the pipeline, the source and destination datasets, and some additional settings of the copy activity, like the retry policy. All these settings can also be changed later from the authoring portal.

The Deployment page opens.

39. Wait for the deployment process to complete. Then click the **Edit Pipeline** button.

> **NOTE** As part of the deployment process, the wizard also runs the pipeline. Just ignore it. You'll re-execute it manually in a moment.

The pipeline canvas opens. Notice that it contains a copy activity.

40. Select the **copy activity**.

The panel below the pipeline canvas changes to contain six tabs. The source of the copy activity is the dataset that points to source JSON files, while the destination is the Parquet dataset. The Mapping tab contains the manual mapping you specified in the Destination Data Store page of the Copy Data wizard.

Next, let's execute the pipeline.

41. Click the **Add Trigger** button and select **Trigger Now** from the drop-down menu that appears.

42. Click **OK** in the Pipeline Run window.

You will be notified when the pipeline has started running and again when the pipeline's execution is complete.

43. To review the pipeline's execution, click the **Monitor** button in the toolbar on the left side of the Azure Data Factory UI panel. Then click the pipeline's name in the list in the Pipeline Runs window.

A window containing details about the pipeline execution opens. It contains a classic pipeline canvas, with green and red icons to show the outcome at a glance. Below the canvas is a list of all activities executed in the last run, along with information like outcome, duration, and so on. You should see just one row related to the copy data activity.

44. Hover your mouse pointer on this row.

Three icons appear: Input, Output, and Details.

45. Click the **Details** icon (it has eyeglasses on it).

A floating pane opens displaying a graphical summary of the copy data execution. It contains important information like the number of records transferred, the total size of the data moved, the number of Data Integration Units used, a breakdown of the total execution time, and so on.

You should see that 211 files have been copied. Notice that the total size of these files in JSON format is around 12 MB, while the compressed Parquet files are just 705 KB. This is a big savings, especially if you consider that these files will be processed in Azure Databricks. In fact, Spark, being an in-memory engine, keeps the footprint of the files to process as low as possible.

46. Use the Azure Storage Explorer desktop or web tool to check that the files have been effectively created. You can find the web version in the Azure Portal in the Storage Explorer (Preview) section of your storage account resource page.

Your storage account should have 211 new files in it under the path /Daily/parquet/ in the staging blob container tree node.

> **NEED MORE REVIEW?** Monitoring Data Factories is a very broad topic and cannot be covered in detail in this book. To learn more about how to visually monitor a Data Factory, see *https://docs.microsoft.com/en-us/azure/data-factory/monitor-visually*. If you prefer to use Azure Monitor to centralize in the monitoring of multiple services and applications, Azure Data Factory has strong integration with it. Read more here: *https://docs.microsoft.com/en-us/azure/data-factory/monitor-using-azure-monitor*.

Create and schedule triggers

Triggers in Azure Data Factory are a convenient way to run pipelines on a specific schedule or in response to an event as soon as it occurs.

A trigger can be bound to multiple pipelines, and a pipeline can have multiple triggers bound to it. The only exceptions are tumbling window triggers, which can be bound to only one pipeline at a time.

Triggers can be created and managed in a dedicated section of the ADF visual authoring portal. As a reminder, you access this portal by clicking the Author & Monitor button in the Azure Portal Data Factory Overview page. (Refer to Figure 3-6.)

The toolbar on the left side of the authoring portal contains a **Manage** button (with a blue toolbox icon). Click this button to open the Data Factory Management panel. Here, you can set up factory-wide settings like linked services, IRs, source control binding, and so on. Click **Triggers** in the **Author** section of the Management panel to open the Triggers page, which lists existing triggers and enables you to create new ones. (See Figure 3-27.)

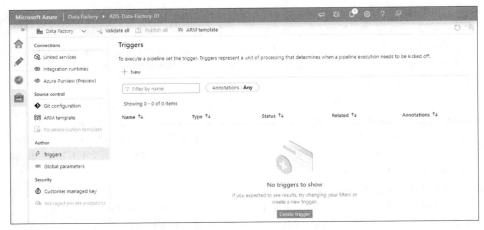

FIGURE 3-27 The Triggers section of the Azure Data Factory UI

Clicking the **+New** button along the top of the Triggers page (or the **Create Trigger** button at the bottom, if the list is empty) opens a New Trigger pane on the right side of the screen. (See Figure 3-28.)

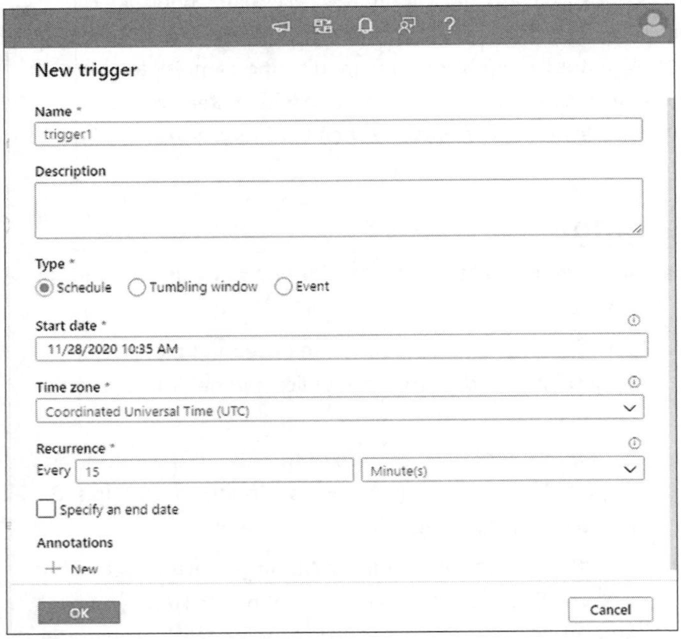

New trigger

Name *
trigger1

Description

Type *
◉ Schedule   ○ Tumbling window   ○ Event

Start date *
11/28/2020 10:35 AM

Time zone *
Coordinated Universal Time (UTC)

Recurrence *
Every 15   Minute(s)

☐ Specify an end date

Annotations
+ New

OK   Cancel

FIGURE 3-28 New Trigger pane

The Name and Description boxes shown in Figure 3-28 are always present in the New Trigger pane. However, the rest of the pane presents different options depending on the value you choose in the Type section because each type of trigger has its own specific properties.

There are three types of triggers:

- **Schedule** Invokes a pipeline on a wall-clock schedule
- **Tumbling window** Operates on a periodic interval, while also retaining state
- **Event** Responds to an event

Schedule triggers

Schedule triggers are probably the most common type of trigger. These have the following parameters:

- **Start Date** The time from when the trigger starts invoking bound pipelines.
- **Time Zone** The reference time zone.
- **Recurrence** The interval between invocations of bound pipelines. Depending on the chosen interval and the value (N) you enter in the Every field, you can set up quite complex schedules:
  - **Minute(s)** The trigger runs every N minute(s).
  - **Hour(s)** The trigger runs every N hour(s).
  - **Day(s)** The trigger runs every N day(s). In addition, you can specify in which hours it will run. For example, you can have a trigger that runs every day at 8 AM, 12:15 PM, 4:30 PM, and 8:45 PM. You can add as many times as needed.
  - **Week(s)** The trigger runs every N week(s). In addition, you can specify in which days of the week and in which hours it will run. For example, you can have a trigger that runs every two weeks on Monday, Wednesday, and Friday, at 5 AM. You can add as many times as needed.
  - **Month(s)** The trigger runs every N month(s). In addition, you can specify on which days of the month (or of the week) and in which hours it will run. Days of the month can be picked directly, or you can select a special Last value to say, for example, "the last day of the month." Days of the week can be paired with First, Second, Third, Fourth, Fifth, and Last selectors to say, for example, "the second Sunday of the month." For instance, you could have a trigger that runs every six months on the first and the third Monday of the month, at 8 AM and 8 PM. You can add as many times as needed.
- **Specify an End Date** You can check this to add a known boundary to the trigger — for example, when you have to stop a schedule as soon as a marketing campaign ends. In this case, you know in advance the end date. This field is optional.
- **Annotations** Key/value pairs that are like tags for resources. These can be useful from a management and monitoring perspective. This field is optional.
- **Activated** Whether the trigger is active. This requires a publish of the Data Factory to take effect. Of course, if you are creating the trigger, a publish is mandatory anyway, but if you change this value after the trigger has been published, a new publish is necessary for this property to take effect.

Tumbling window triggers

Tumbling window triggers fire at a periodic time interval from a specified start time while retaining state. This type of trigger uses a fixed, non-overlapping window (with a size equal to the specified interval) to invoke the pipeline bound to it. Say, for example, that you have a backfill scenario: an archive of daily files that you want to import. The first file is dated to January 1, 2015. You can define a tumbling window trigger with a start date equal to the older file and an interval of 24 hours. As soon as you activate the trigger, it will begin executing its bound pipeline with a start date of January 1, 2015. It will then invoke another execution with a start date of January 2, 2015, and so on, until it reaches the defined end date. If no end date has been specified, the trigger will keep invoking the pipeline at specified intervals.

With tumbling window triggers, you can also specify a degree of parallelism higher than 1 to have parallel executions of the pipeline to speed up the process. The pipeline, through parameters, can access the WindowStart and WindowEnd system properties of the trigger instance and use them to configure its execution at runtime. For example, it can search for a file named 2015-01-01.txt in the first execution, 2015-01-02.txt in the second execution, and so on.

A tumbling window trigger has the following parameters:

- **Start Date** This is the time from when the trigger starts invoking the bound pipeline. Since there is no Time Zone selector for this type of trigger, the start date has to be specified in UTC.

- **Recurrence** This is the interval between invocations of the bound pipeline or, in other words, the size of the non-overlapping window. Depending on the chosen interval and the value (N) you enter for the Every field, you can set up the following schedules:

  - **Minute(s)** The window size is equal to N minute(s).

  - **Hour(s)** The window size is equal to N hour(s).

- **Specify an End Date** This can be checked to add a known boundary to the trigger — for example, when you have to stop a schedule as soon as a marketing campaign ends. In this case, you know in advance the end date.

- **Annotations** Key/value pairs that are like tags for resources. These can be useful from a management and monitoring perspective.

- **Activated** Whether the trigger is active. This requires a publish of the Data Factory to take effect. Of course, if you are creating the trigger, a publish is mandatory anyway, but if you change this value after the trigger has been published, a new publish is necessary for this property to take effect.

In addition, the Advanced section of the New Trigger panel for tumbling window triggers has the following properties:

- **Add Dependencies** You can make this trigger dependent on another tumbling window trigger in your Data Factory, ensuring that your trigger will run only after a successful execution of the trigger it depends on.
- **Delay** The delay between the processing of windows.
- **Max Concurrency** The number of simultaneous trigger runs that are fired.
- **Retry Policy: Count** The number of retries before a pipeline run is marked as failed.
- **Retry Policy: Interval in Seconds** The delay between each retry attempt.

Event triggers

An event trigger is fired in response to a specific event, such as the creation or the deletion of a file on an Azure Blob storage account. It has the following parameters:

- **Storage Account** The storage account to listen to for events. At the time of writing, only ADLS Gen2 and General Purpose version 2 storage accounts are supported. You can select the account from the ones available in your visible subscriptions or you can enter the information needed to connect to it manually.
- **Container Name** The container of your storage account to monitor for events.
- **Blob Path Begins With** The first characters of the path to monitor for events.
- **Blob Path Ends With** The last characters of the path to monitor for events.
- **Event** The events that will trigger an execution. They can be blob created, blob deleted, or both.
- **Ignore Empty Blob** Indicates whether an event related to files of 0 bytes will fire the trigger.
- **Annotations** Key/value pairs that are like tags for resources. These can be useful from a management and monitoring perspective. This field is optional.
- **Activated** Whether the trigger is active. This requires a publish of the Data Factory to take effect. Of course, if you are creating the trigger, a publish is mandatory anyway, but if you change this value after the trigger has been published, a new publish is necessary for this property to take effect.

Clicking the Continue button at the bottom of the New Trigger pane opens a data preview screen. In this screen, you can see which blobs could be potentially affected by your parameter selections to see if they will work well or are too broad.

Attach a trigger to a pipeline

To attach a trigger to a pipeline, open the pipeline in the editor and click the **Add Trigger** button along the top. Then select a previously created trigger in the right panel that opens (see Figure 3-29) or create a new one directly from there.

FIGURE 3-29 Add triggers to a pipeline

In addition to using triggers, you can invoke pipelines manually in the authoring portal (as you did in the "Use the Copy Data wizard to create and run a simple pipeline" practice exercise) or through a rich set of APIs, including the following:

- .NET SDK
- Azure PowerShell module
- REST API
- Python SDK

NEED MORE REVIEW? To learn more about triggers in Data Factory and all the available options to invoke a pipeline execution, see *https://docs.microsoft.com/en-us/azure/data-factory/concepts-pipeline-execution-triggers.*

Implement Azure Databricks clusters, notebooks, jobs, and autoscaling

Before you can execute any work in Azure Databricks, you have to set up a cluster. You can create a cluster graphically in a dedicated section of the Azure Databricks workspace. As a reminder, to access the workspace, go to your resource's Overview page and click the **Launch Workspace** button. (Refer to Figure 3-11.)

Once in your workspace, click the **Clusters** button in the toolbar on the left to open the Cluster Management page. This offers an overview of existing clusters and their status.

To create a new cluster, click the **+Create Cluster** button. This opens the Create Cluster page. (See Figure 3-30.) Here, you define the following parameters:

- **Cluster Name** A friendly name, this must be unique within the workspace.
- **Cluster Mode** Options are Standard (single-user oriented) and High Concurrency (more suitable for parallel workloads).
- **Pool** Specifies whether the cluster should be added to a serverless pool or be a stand-alone one. If you add it to an existing pool, the Databricks runtime handles it for you, using it or not depending on the workload and the pool configuration.

- **Databricks Runtime Version**   The runtime your cluster will run. This is a combination of Scala and Spark versions, support for GPU acceleration (very suitable for AI algorithms, particularly neural networks and deep learning), and optimization for ML development. You can choose between current, older, and beta releases.

- **Autopilot Options**   Enable (or not) auto-scaling and auto-shutdown. For the latter, you can specify the inactivity timeout period (in minutes) before a node should be shut down.

- **Worker Type and Driver Type**   Define the size of the nodes of your cluster. You can pick from a selection of Azure virtual machine series. In addition, you can specify the number of worker nodes (if auto-scaling is disabled) or a minimum/maximum range of worker nodes the runtime will throttle between (if auto-scaling is enabled).

- **Advanced Options**   Fine-tune your cluster. You can set Spark properties to override the default configuration, add environment variables, define a custom path for log files, and specify the path to init scripts the cluster has to run on every node when provisioning them (for example, to install custom libraries). In addition, you can check the Azure Data Lake Storage Credential Passthrough option, which automatically passes the AD credentials of a specific user (if using Standard cluster mode) or of the current user (if using High Concurrency cluster mode) when reading from or writing data to an ADLS. Both Gen1 and Gen2 are supported.

FIGURE 3-30   The New Cluster page

To provision a cluster, click the **Create Cluster** button at the top of the page. As soon as the provision is complete, the cluster is ready to accept Spark jobs.

If your workspace is on the Premium pricing tier, you can use role-based access control (RBAC) to secure access to your clusters. Thanks to the integration between Azure Databricks and Azure Active Directory (AD), you can centralize the management of users in AD without the need to replicate logins in Azure Databricks. Groups defined in Azure AD are not visible in the workspace, but you have the option to create local groups to avoid assigning rights user by user.

You can author Spark code in notebooks, which are briefly described and used in the section "Ingest data into Azure Databricks" later in this chapter. As noted, notebooks can be executed manually, via REST API, or under schedule.

You can schedule a notebook with an external orchestrator, like Azure Data Factory (the preferred option for complex workloads), or through the internal job scheduler, which you can access by clicking the **Jobs** button in the left toolbar. Clicking the **+Create Job** button at the top of the page opens the job creation page shown in Figure 3-31. On this page you can define the following properties:

- **Name**   The name of the job.
- **Task**   The task that will be performed by the job. This could be an existing notebook, a compiled JAR, or a spark-submit command that invokes a JAR or a script.
- **Cluster**   The cluster that will be used to execute the job. You can have clusters that are dedicated only to executing jobs. These are called *job clusters*. The cost of this type of cluster is computed differently from clusters meant to be used for development or interactive execution, called *all-purpose clusters*.
- **Schedule**   The schedule of the job, based on minutes, hours, days, or months. Also, you can use Cron syntax, which is very familiar to Unix users.
- **Alerts**   Email notifications for events such as job start, job success, or job failure.
- **Maximum Concurrent Runs**   Specifies whether concurrent runs are allowed, and if so, how many.
- **Timeout**   Specifies how long (in minutes) the job can run before it gets killed.
- **Retries**   The retry policy — the number of retry attempts, the delay between retries, and whether to do a retry after a timeout.
- **Permissions**   Defines which users and groups can view the job and manage it.

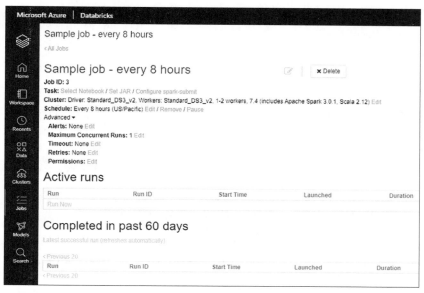

FIGURE 3-31 The job creation screen

In addition, below the job properties are two lists: Active Runs and Completed in Past 60 Days.

PRACTICE   Provision an Azure Databricks workspace, create a cluster, and mount an external storage

This practice exercise guides you through the creation of an Azure Databricks workspace, the setup of a cluster inside it, and the mounting of an external storage to that cluster. The following section, "Ingest data into Azure Databricks," uses these resources to show a simple data-engineering workload.

> **NOTE**   You need a pay-as-you-go subscription to complete this practice exercise because the quota limits for the free trial subscription prevent you from creating even a bare-minimum cluster. Specifically, vCores are limited to four, which equals the minimum number of cores for worker and driver nodes you can select when creating a cluster — meaning that the simplest cluster you can create, composed of the driver node and a single worker node, would require eight cores. (The option to increase quota limits by issuing a ticket to Azure Support is not available for free trial subscriptions.)

1. Open *https://portal.azure.com* and use your credentials to log in.

2. Click **+Create a Resource**.

3. Type **Azure Databricks** in the search box and select the corresponding entry that appears in the drop-down list.

4. Click **Create**.

5. In the **Azure Databricks** page, in the **Basics** tab, configure the following settings:

   ■ **Subscription**   Choose the subscription that will contain your Azure Databricks workspace.

   ■ **Resource Group**   Choose the resource group for your Azure Databricks workspace.

   > **NOTE**   If the resource group does not exist yet, you can create it from here.

   ■ **Region**   Choose the same region you used in previous practices and exercises (for example, West Europe).

   ■ **Name**   Type a name for the Azure Databricks workspace. This must be globally unique.

   ■ **Pricing Tier**   Select Trial.

6. Skip the **Networking** tab.

7. Click **Review +Create**.

8. If your new **Azure Databricks** workspace is successfully validated, click **Create**.

9. After the Azure Databricks workspace is provisioned, navigate to its Overview page and click the **Launch Workspace** button.

   The control plane web application opens, where you can interact with your Azure Databricks instance.

10. Click the **Clusters** button in the left toolbar.

11. Click the **+Create Cluster** button.

    The New Cluster page opens.

12. Enter a friendly name for the cluster in the **Cluster Name** box.

13. Open the **Cluster Mode** drop-down list and choose **Standard**.

14. Open the **Pool** drop-down list and choose **None**.

15. In the **Databricks Runtime Version** drop-down list, choose the latest stable (not Beta) runtime, but avoid runtimes containing the ML suffix. For example, choose a runtime like **7.4 (Scala 2.12, Spark 3.0.1)**.

16. Select the **Enable Autoscaling** option check box.

17. Optionally, set a different **Terminate After *x* Minutes of Inactivity** value for cluster termination. The default is 120 minutes.

18. Open the **Worker Type** drop-down list and choose **Standard_DS3_v2**.

19. Type **1** in the **Min Workers** box and type **2** in the **Max Workers** box.

20. Open the **Driver Type** drop-down list and select **Same as Worker**.

21. Click the **Create Cluster** button at the top of the page.

    The cluster-creation process might take a few minutes. When the process is complete, you will be redirected to the cluster edit page. A green circle next to the name of the cluster at the top of the page indicates that it is running.

22. Click the **Home** button. The Workspace panel appears on the left.

23. In the Workspace panel, click the **down arrow** next to the folder with your username (it has a small house icon next to it) and choose **Import** from the menu that opens.

24. Click the **Browse** link. Then locate and select the **ADS.dbc** file in the companion content.

    Azure Databricks allows you to export single notebooks or to export an entire folder structure in various formats. DBC, which is essentially an archive file in a proprietary format, is one of them. The ADS.dbc file contains a Databricks folder with some notebooks you will use in this and other practice exercises.

25. Click the **Import** button.

26. Click the **ADS folder** you just imported and select the **00 - DBFS** notebook inside it.

27. In the **00 - DBFS** notebook, fill in the following missing variables in the **Cmd 2** cell as follows:

    - **<container-name>**
    - **<account-name>**
    - **<access-key>**

    For the first two variables, you can use the same account and source container you used in the "Use the Copy Data wizard to create and run a simple pipeline" practice exercise earlier in this chapter. For the last variable, you can find the storage account's access key in the Access Keys section of the account's page on the Azure Portal. (You can choose either the primary or the secondary key.)

28. Hover your mouse pointer over the **Cmd 2** cell, click the **Run** icon in the top-right menu that appears, and choose **Run Cell** in the drop-down list that opens. Alternatively, press **Ctrl+Enter** to run the currently highlighted cell. Then wait for the run to complete.

    If the notebook has not already been attached to a cluster (using the drop-down list along the top), you will be prompted to attach it at first run. In addition, if the cluster is not running, the engine will attempt to start it for you. Alternatively, you can start it manually in the Cluster Management section of the UI before running the cell. If this is the first time you have run this cell, you might get a `java.rmi.RemoteException`. In fact, because the container has yet to be mounted to the cluster, the unmount operations in the second row will fail.

29. Fill in the following missing variables in the **Cmd 3** cell:

   - **<container-name>**
   - **<account-name>**
   - **<access-key>**

   For the first two variables, you can use the same account and destination container you used in the "Use the Copy Data wizard to create and run a simple pipeline" practice exercise earlier in this chapter. For the last variable, you can find the storage account's access key in the Access Keys section of the account's page on the Azure Portal. (You can choose either the primary or the secondary key.)

30. Hover your mouse pointer over the **Cmd 3** cell, click the **Run** icon in the top-right menu that appears, and choose **Run Cell** in the drop-down menu that opens. Alternatively, press **Ctrl+Enter** to run the currently highlighted cell. Then wait for the run to complete.

   If this is the first time you have run this cell, you might get a `java.rmi.RemoteException`. In fact, because the container has yet to be mounted to the cluster, the `unmount` operations in the second row will fail.

31. Hover your mouse pointer over the **Cmd 5** cell, click the **Run** icon in the top-right menu that appears, and choose **Run Cell** in the drop-down menu that opens. Then wait for completion.

   Just below the box containing the code, you will see a grid view listing the content of the container you just mounted.

32. Repeat this same step for the **Cmd 6** cell.

Ingest data into Azure Databricks

This section takes a closer look at Spark from a development point of view. In fact, you will use JSON and Parquet files as source data to see how to actually perform data transformation with the Spark engine.

The access point to the data is the data frame object, which is a layer over the data and can be instantiated in many ways. Think of it as an external table in Hive or PolyBase, but more complex and powerful.

Once created, a data frame is immutable. In fact, performing any transformation on it produces another data frame as output, which maintains the lineage with its parent. Transformations include operations like filtering, grouping, projecting, and so on. When you call for an action on the data frame, like displaying some records on the UI or writing data to disk, the Spark engine runs across its lineage to track down all the transformations it has to apply before returning the data. Then, the engine produces the physical plan for the job and submits it to the executors, which start collecting the data from the data store and follow the given instructions.

Obviously, the more transformations that have been chained before calling the action, the more difficult it will be for the engine to find the most optimal plan. For this reason, in many cases, it is better to break the chain of transformations into smaller parts, writing intermediate results to disk and rereading them right after creating a fresh data frame.

Spark makes some tasks very easy, like working with files stored on disk. For example, once you mount a storage account to the cluster, you can instantiate a data frame that points to the COVID-19 Parquet data prepared in the "Use the Copy Data wizard to create and run a simple pipeline" practice exercise earlier in this chapter by using one of the syntaxes shown in Listing 3-1. In this case, the PySpark language has been used.

> **NOTE**   You can find the full code shown here in the 01 – Transform data notebook inside the ADS.dbc archive in the companion content. If you did the "Provision an Azure Databricks workspace" practice exercise earlier in this chapter, this notebook should already be available in your workspace.

LISTING 3-1   Data frame creation in PySpark

```
# longer version
#df = spark.read.format('parquet').load('/mnt/ads-staging/Daily/parquet');

# shorter version
df = spark.read.parquet('/mnt/ads-staging/Daily/parquet');
```

In case you are new to Spark, here are a few points to keep in mind:

- To interact with a Spark cluster, you need to create a session object — more precisely, a SparkSession object. Databricks creates this for you behind the scenes, and the word spark in the code is a shorthand for it.

- The read method of the SparkSession class returns a DataFrameReader object, which in turn can be used to return a data frame. This allows for some format specifications, like the file schema, whether the source file has a header, and so on.

- The path points to a folder, and Spark can point to the entirety of the folder's content with the same data frame. (Think of it like a merge of all the files in the folder.) You can use wildcards to select some portion of the folder content (for example, *.txt). As you saw in the "Provision an Azure Databricks workspace, create a cluster, and mount an external storage" practice exercise earlier in this chapter, containers accessed in Listing 3-1 to Listing 3-7 have been mounted to the cluster using the code contained in cells Cmd 2 and Cmd 3 of the 00 - DBFS notebook.

- The two different versions shown in Listing 3-1 are exactly equivalent in terms of performance. In fact, they produce the same code. Most popular formats, like CSV, JSON, Parquet, and more, have a dedicated method in the DataFrameReader class with format-specific optional parameters, but you can always use the more generic syntax format('<supported format>').load('<path>').

> **NEED MORE REVIEW?**   You can find out more about the DataFrameReader class in the Apache Spark documentation: *https://spark.apache.org/docs/latest/api/java/org/apache/spark/sql/DataFrameReader.html*.

Figure 3-32 shows the output of the code in Listing 3-1. Because the engine assembled information about the column names and numbers and nothing more, the command executes very quickly. Also, the data frame schema is exactly the one you specified in the copy activity in Azure Data Factory. In fact, Parquet files include information about their columns' data types, something that CSV and JSON formats cannot do.

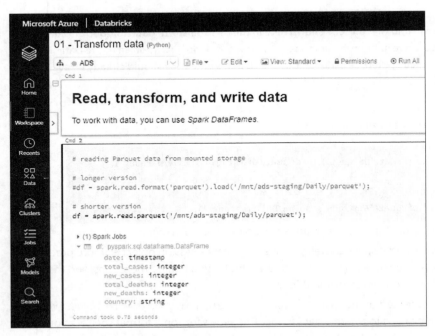

FIGURE 3-32 Data frame creation output

If you are reading from a schema-less file, Spark will return by default a data frame with all columns set to a string data type. If you want to specify a schema, you have two options:

1. **The inferSchema option**   If you set this property to True when you create the data frame, you instruct the engine to automatically deduce the schema from the data. This is a very handy but dangerous option, because Spark performs a full scan of the source file(s) to understand its schema at the time of data-frame creation, issuing a specific job for this task. If you are working with large (or unknown in advance) volumes of data, this option is not advisable. In addition, the schema may change unexpectedly on subsequent executions. In fact, as data changes or new data comes in, it may contain dirty records, or simply more complete ranges of values. As a result, a column that always contained numbers could suddenly present alphanumeric characters in some fields. This option is always on when reading JSON files and cannot be specified.

2. **Providing a schema** In this case, you create and populate a StructType object containing the expected fields and their data type. This object is then passed to DataFrameReader to instruct it about the schema of the file(s) it has to read.

Listing 3-2 shows both approaches, reading from daily files in JSON format instead of Parquet ones. Notice that the syntax used to read from JSON files is very similar to the one used in Listing 3-1 for Parquet files.

LISTING 3-2 File schema in PySpark

```
# 1. Inferred schema
df = spark.read.json('/mnt/ads-globaldata/Daily');

# 2. Explicit schema
from pyspark.sql.types import *;

fileSchema = StructType([
  StructField("Code", StringType()),
  StructField("Country", StringType()),
  StructField("date", StringType()),
  StructField("new_cases", LongType()),
  StructField("new_cases_per_million", DoubleType()),
  StructField("new_cases_smoothed", DoubleType()),
  StructField("new_cases_smoothed_per_million", DoubleType()),
  StructField("new_deaths", LongType()),
  StructField("new_deaths_per_million", DoubleType()),
  StructField("new_deaths_smoothed", DoubleType()),
  StructField("new_deaths_smoothed_per_million", DoubleType()),
  StructField("stringency_index", DoubleType()),
  StructField("total_cases", LongType()),
  StructField("total_cases_per_million", DoubleType()),
  StructField("total_deaths", LongType()),
  StructField("total_deaths_per_million", DoubleType())
]);

df = spark.read.json('/mnt/ads-globaldata/Daily', schema = fileSchema);
```

Figure 3-33 displays the results of both executions. You can see that the first approach took 3.84 seconds, while the second approach took 0.26 second. To infer a file schema, Spark has to read the whole content of the folder; the number of Spark jobs issued (three) is evidence of that. Be aware that this gap increases as the data volume grows.

FIGURE 3-33 Inferred and explicit schema comparison

To peek at the file content, you can use the show action in the data frame, which displays the records in plain text. Or, for a fancier result grid, you can use the Databricks display method, which outputs the first 1,000 records of the data frame in an interactive table. (See Figure 3-34.) The table can be ordered, exported, and transformed into various chart types to see the content of the data frame in a graphical way.

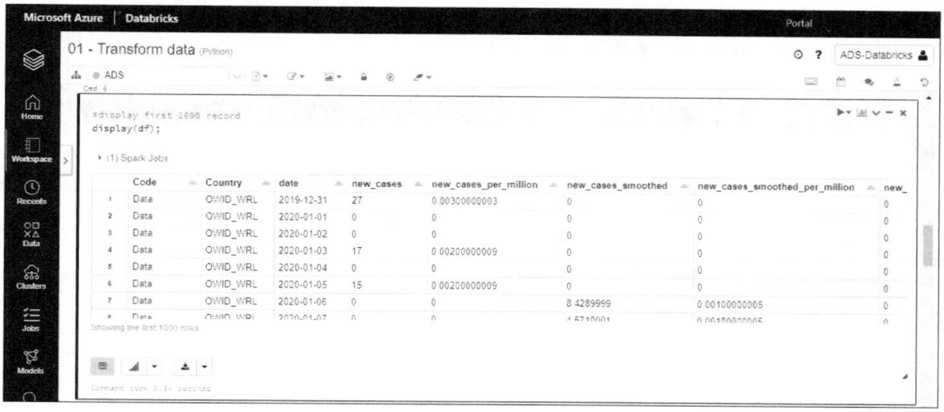

FIGURE 3-34 Output of the `display` command

Suppose you need to extend data about COVID-19 cases with country information, but only for countries in North America and South America. This information is stored in the covid-data_Countries.json file you uploaded to your storage account along with the daily data files. In addition to that, suppose you want to add a moving average of new cases reported over the last seven days. The resulting dataset has to be written to a storage account in CSV format and must include column headers.

To produce the desired output, the first step is to instantiate two data frames — one for the daily data and one for the demographic (including country) information. (See Listing 3-3.)

LISTING 3-3   Reading the source data in PySpark

```
dfData = spark.read.parquet('/mnt/ads-staging/Daily/parquet');
dfCountries = spark.read.json('/mnt/ads-globaldata/Demography');
```

Then, if you register each data frame as a temporary table, you can use a familiar SQL syntax to join them using the `sql` command of the `SparkSession` object or writing SparkSQL code.

A data frame object also has transformations like `join`, `groupBy`, `select`, and others, which can be used to manipulate data programmatically. But one nice thing about Spark is that you can approach different types of work with the semantics you like the most. Listing 3-4 displays the code needed to register data frames as temporary tables (or views).

LISTING 3-4   Create temporary tables in PySpark

```
dfData.createOrReplaceTempView('vwData');
dfCountries.createOrReplaceTempView('vwCountries');
```

Before creating the output data frame, you want to leverage the temporary tables you just created to test the moving average and display the output in a graphical way. Using the `%sql` directive in a cell, you can write SparkSQL code directly, as shown in Figure 3-35. As you can see, the resulting dataset is displayed in a line chart. Notice that to compute the moving average, you can choose from a very comprehensive set of window functions. In this case, we used an average aggregation over a fixed seven-day window. (We have assumed for the sake of simplicity that no gaps exist between reported dates.)

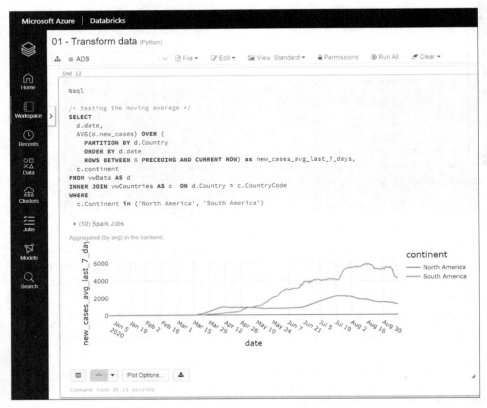

FIGURE 3-35 SparkSQL code

When you are satisfied with the results, create the output data frame with the code in Listing 3-5. This code executes almost instantly because it does not include actions. It simply stacks some transformations over the data frames instantiated in Listing 3-3.

LISTING 3-5 Output data frame creation in PySpark

```
dfExtendedData = spark.sql('''
SELECT
  d.date,
  d.total_cases,
  d.new_cases,
  d.total_deaths,
```

```
    d.new_deaths,
    AVG(d.new_cases) OVER (
      PARTITION BY d.Country
      ORDER BY d.date
      ROWS BETWEEN 6 PRECEDING AND CURRENT ROW) as new_cases_avg_last_7_days,
    c.aged_65_older,
    c.aged_70_older,
    c.life_expectancy,
    c.population,
    c.population_density,
    c.location,
    c.continent
FROM vwData AS d
INNER JOIN vwCountries AS c  ON d.Country = c.CountryCode
WHERE
    c.Continent in ('North America', 'South America')
'''');
```

To write the results to disk, use the data frame's write action, specifying the destination path. (See Listing 3-6.)

LISTING 3-6   Writing data in PySpark

```
dfExtendedData.write.mode("overwrite").csv('/mnt/ads-staging/transform/covid19_data',
header = True);
```

This points to an output folder (and not to a file) because by default the engine could divide it into chunks, depending on the final number of data partitions Spark ends up with after all the transformations. Although you cannot change the output file name with the DataFrame API, you can at least ensure you obtain a single file as a result by using the coalesce action. This narrows down the number of data-frame partitions to a specified parameter. (See Listing 3-7.) Figure 3-36 displays the output folder and the file content.

LISTING 3-7   Writing data to a single file in PySpark

```
dfExtendedData.coalesce(1).write.mode("overwrite").csv('/mnt/ads-staging/transform/covid19_
data_single_file', header = True);
```

> **NOTE** Before changing the engine behavior, it is important to understand what is happening behind the scenes. A good starting point is this page on performance tuning in the Apache Spark documentation: *https://spark.apache.org/docs/latest/sql-performance-tuning.html.*

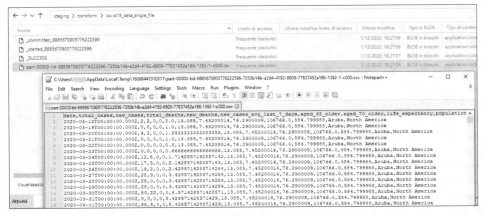

FIGURE 3-36 The content of the covid19_data_single_file directory

Ingest and process data using Azure Synapse Analytics

The final target of a batch workload in a modern data warehouse scenario is typically a high-performing data warehouse, capable of storing huge volumes of data in a scalable way and accessing it very fast. In Azure, this usually translates to Azure Synapse Analytics.

As explained in Chapter 2, when you provision an Azure Synapse Analytics resource in your subscription, you get a workspace. This workspace can contain three types of data services:

- **SQL serverless pool** This is a ready-to-go SQL-based pool. You can have exactly one pool of this kind in your workspace. It comes very handy for exploring data with little or no configuration. This pool is billed on a consumption model, so you do not pay for *how long* you use it, but rather for *how much data* you move. Its typical use cases are non–mission critical data exploration, data conversion, and transformation workloads.

- **SQL dedicated pool** This pool was formerly known as Azure SQL Data Warehouse. In fact, it is exactly that. When you provision a SQL dedicated pool, you create a fully functional Azure SQL Data Warehouse with all the features that made that service a premium choice for high-performance data warehouses. Its typical use cases are data warehouse relational storage and mission-critical ELT workloads.

- **Serverless Spark pool** This pool allows you to run Spark-based workloads in your workspace. The Spark engine behind it is a Microsoft-proprietary version of the open-source Spark project. Currently, it supports the Scala, PySpark, and C# languages. The latter has been made possible by the Spark.NET open-source project from Microsoft. Its typical use cases are data exploration, data conversion, stream processing, and mission-critical data engineering and ML workloads.

Because these pools expose an endpoint, issuing queries and running scripts on them can be made with many traditional tools, like Microsoft SQL Server Management Studio, Microsoft Visual Studio, and Visual Studio Code. However, Spark workloads are not supported by some of them. For example, you cannot use Microsoft SQL Server Management Studio to run Spark jobs against a Spark pool; instead, you should use Visual Studio Code.

Azure Synapse Studio is another way to interact with these pools. It is a very comprehensive web application, as shown in Figure 3-37. You already used it in the practice exercises in Chapter 2 when you created a dedicated SQL pool and a Spark pool.

FIGURE 3-37 Azure Synapse Studio home page

The main pane contains some quick links to typical actions you can perform in your workspace:

- **Ingest**   Opens the Copy Data wizard, which is nearly identical to the tool of the same name in Azure Data Factory, which you used in the "Use the Copy Data wizard to create and run a simple pipeline" practice exercise earlier in this chapter.
- **Explore and Analyze**   Leads to a quick start guide about data exploration on data lakes.
- **Visualize**   Opens a pane where you can connect a Power BI workspace. After you set up this connection, you can start authoring BI reports directly inside Azure Synapse Studio.
- **Learn**   Opens the Knowledge Center, a repository of tutorials and samples of features and capabilities of Azure Synapse Analytics.

The left panel allows you to reach various Azure Synapse Studio pages:

- **Data**   Manage and explore data in your databases and linked services. Databases can be created on dedicated SQL pools as well as on serverless SQL pools or Spark pools. In the last two cases, databases are usually a repository for metadata, and contain external tables pointing to blob storage files and folders.
- **Develop**   Create SQL scripts, notebooks (for authoring Spark code), and data flows. The latter behave as in Azure Data Factory, allowing for no-code data transformation backed by on-demand Azure Databricks clusters. (At the time of writing, you cannot use a Spark pool created in your workspace to execute a data flow.) Scripts and notebooks have IntelliSense support, can be organized in folders, and support collaborative online code editing from multiple users.
- **Integrate**   Create visual integration pipelines, like you do in Azure Data Factory. It also offers dedicated tasks, like executing a Synapse-type notebook against a Spark pool.

- **Monitor** Monitor the health state of the resources in your workspace. You can also check all your pipeline runs and even expand every single task to track down performance issues.

- **Manage** Create and manage pools, triggers, security, and source control binding. You can also add linked services to integrate external resources with your workspace in the same way you do in Azure Data Factory. For example, you can connect storage accounts, Power BI workspaces, and Azure ML workspaces. Connecting storage accounts makes it very easy for users and developers to explore data contained in them and set up quick transformation or data-movement workloads.

Data ingestion and processing can follow different paths, depending on the type of pool you target and the approach you want to take. For easier access to your data, it is always a good thing to add a linked service that points to the storage account where your data resides. Figure 3-38 shows the ADLS Gen2 New Linked Service tab. You can access it from the Manage area in the Linked Services section. If you click the **Add** button along the top and choose a linked service of the desired type, you'll see a very familiar tab on the right, which asks for the same information and parameters that linked service in Azure Data Factory.

FIGURE 3-38 The New Linked Service tab

Working with serverless and dedicated SQL pools

The two most common approaches to working with external data in SQL pools are:

- External tables
- The OPENROWSET statement

EXTERNAL TABLES

External tables are usually the preferred way to map external data because they enable cross-pool references. For example, you can create an external table in a Spark pool, and then have users read its content with SQL commands issued against a serverless SQL pool. This way, you can decide which engine best fits a particular workload without compromising between performance and accessibility.

To read from and write to external tables in serverless and dedicated SQL pools, you can leverage the PolyBase component. PolyBase is central not only for data transformation, but also for the *L* part of an ETL/ELT process (the load phase). Also, PolyBase is a performance-wise approach to ingest data that resides in blob storage into traditional tables in a dedicated SQL pool.

> **NOTE** The COPY statement, which recently reached general availability, allows for an easier approach to data movements in and out of Azure Synapse Analytics. Read more here: *https://docs.microsoft.com/en-us/sql/t-sql/statements/copy-into-transact-sql?view=azure-sqldw-latest*.

The version of PolyBase that ships with Azure Synapse Analytics fully supports Azure Blob storage and ADLS, both for the import and export of data.

> **NOTE** The procedure is described in detail in the "Implement PolyBase" section in Chapter 2. In addition, you can find just the code needed to set up PolyBase in Azure Synapse Analytics in the synapse_configure_polybase.sql file in the companion content.
>
> This procedure applies to dedicated SQL pools in Azure Synapse, but there are some differences when targeting a serverless SQL pool. Read more at *https://docs.microsoft.com/en-us/sql/t-sql/statements/create-external-data-source-transact-sql*.

After you set up PolyBase, you create an external table that points to the source file. Listing 3-8 shows an example of external table creation. The full code can be found in the synapse_create_database_objects.sql file in the companion content; in addition, there is a similar procedure in the SQL Database.ipynb file, which you used in the "Implement relational data stores" section in Chapter 2 to map COVID country information.

LISTING 3-8 Creating an external table in Azure Synapse Analytics

```
-- Create the external table
CREATE EXTERNAL TABLE dbo.covid_data (
  [date] NVARCHAR(255),
  total_cases INT,
  new_cases INT,
  total_deaths INT,
  new_deaths INT,
  new_cases_avg_last_7_days NVARCHAR(255),
  aged_65_older NVARCHAR(255),
  aged_70_older NVARCHAR(255),
  life_expectancy NVARCHAR(255),
  population NVARCHAR(255),
  population_density NVARCHAR(255),
```

```
   location NVARCHAR(255),
   continent NVARCHAR(255)
)
WITH (
    LOCATION = '/transform/covid19_data_single_file',
    DATA_SOURCE = <datasource_name_here>,
    FILE_FORMAT = CsvFileFormat
);

CREATE STATISTICS stats_covid_data_01 ON covid_data(location);
```

The WITH clause contains all the information needed to correctly read the file: where it is (LOCATION), how to access it (DATA_SOURCE), and how to interpret its content (FILE_FORMAT).

You can now load data into Azure Synapse Analytics, as shown in Listing 3-9.

LISTING 3-9 Loading data into Azure Synapse Analytics

```
BEGIN TRY
  DROP TABLE dbo.covid_data_staging;
END TRY
BEGIN CATCH
END CATCH;

CREATE TABLE dbo.covid_data_staging
WITH
(
  CLUSTERED COLUMNSTORE INDEX,
  DISTRIBUTION = HASH(location)
)
AS
SELECT
   *
FROM dbo.covid_data;
```

It is not recommended to load data directly to the final table. Instead, use a staging table that acts as a bridge. This is mainly because the final table usually has a physical structure that is well-suited for fast data retrieval (for example, it has one or more indexes), but not necessarily for data insert.

Loading data to a table with no indexes (technically called a *heap*) is a well-known best practice in data warehousing. It is a very effective way to keep the data transfer very short. In fact, decoupling external sources from your engine should be a very high priority. You can perform additional transformations on the staging table and, when you are finished, you can change its physical form to mimic the destination table if needed — for example, with techniques like partition switching (when applicable).

> **NOTE** If you use a partition-switching pattern, it could be that your staging table was already partition-aligned with your destination at the time of insertion and is not a heap.

NOTE You can still use tools and programmatic approaches like bcp and the SqlBulkCopy API to load data into Azure Synapse Analytics. They are slower than PolyBase and the COPY statement, though, and for this reason they are not the preferred method.

The creation of the local staging table should always be done through the CREATE TABLE AS SELECT (CTAS) statement. This statement performs a parallel operation that creates and fills the table from the output of a SELECT with a single command. This is very important because it leverages the scalable nature of Azure Blob storage, granting optimal performance when reading data.

NOTE To learn more about the CREATE TABLE AS SELECT statement, see *https://docs.microsoft.com/en-us/azure/synapse-analytics/sql-data-warehouse/sql-data-warehouse-develop-ctas*.

If you are using the Azure Synapse Analytics engine for data transformation (and not data loading), your target would be probably a file on Azure Blob storage instead of a table. In this case, you must use a slightly different statement: CREATE EXTERNAL TABLE AS SELECT (CETAS). You can find additional information about such statements here: *https://docs.microsoft.com/it-it/azure/synapse-analytics/sql/develop-tables-cetas*.

Another important aspect worth mentioning is the elasticity of the service. Taking advantage of Azure Synapse Analytics' ability to scale out can help you achieve the best performance possible and, at the same time, keep costs reasonable. It is very common to increase the compute firepower just for the load and transform phases and decrease it afterward.

NEED MORE REVIEW? For a detailed overview of data-loading patterns and best practices for Azure Synapse Analytics, see *https://docs.microsoft.com/en-us/azure/synapse-analytics/sql-data-warehouse/design-elt-data-loading*.

The copy activity in Azure Data Factory and the sink transformation in ADF mapping data flows make it very easy to leverage PolyBase when using Azure Synapse Analytics as a sink. All you have to do is select the **Enable Staging** property in the **Activity** settings and specify a storage account to hold the data mapped by the external table.

NEED MORE REVIEW? Azure Databricks has an optimized connector for Azure Synapse Analytics, too. A detailed post can be found in the documentation here: *https://docs.databricks.com/data/data-sources/azure/synapse-analytics.html*.

THE OPENROWSET COMMAND

The OPENROWSET command offers a very flexible way to read data that resides in blob storage when targeting a serverless SQL pool. It supports BULK operations, which means data can be read in an optimized, and therefore fast, way rather than row-by-row. Also, it has schema-inferring capabilities, so you need not specify the file schema before reading its content. This is great for data exploration, where you usually want to get a sneak peek of the data as soon as possible.

Supported formats are as follows:

- **CSV** Includes any delimited text file with row/column separators. Separators can be customized, so for example, you can specify tab to parse TSV files. If no schema is specified when reading data, the engine tries to determine the columns that compose the file and their data types.

- **Parquet** Includes the widely used binary Parquet format. You can also specify a subset of columns you want to be returned rather than the full list. Parquet files already contain information about columns and their data types.

Listing 3-10 shows a simple use of the OPENROWSET command, which allows for the reading of the same content mapped in Listing 3-8 using PolyBase. As you can see, this code lets the engine infer the schema of the file because it does not specify any structure for it.

LISTING 3-10 Using the OPENROWSET command

```
SELECT
  t.*
FROM
  OPENROWSET(
    BULK 'https://adsstoragedatalake.dfs.core.windows.net/staging/transform/covid19_data_
single_file/*.csv',
    FORMAT = 'CSV',
    PARSER_VERSION = '2.0',
    HEADER_ROW = TRUE
  ) as t
```

> **NEED MORE REVIEW?**   The engine is accessing the file using the credentials of the user executing the query, leveraging Azure Active Directory credential pass-through. This is not the only option available, of course. Read more here: *https://docs.microsoft.com/en-us/azure/synapse-analytics/sql/develop-openrowset#security.*

The query in Listing 3-10 produces the output shown in Figure 3-39.

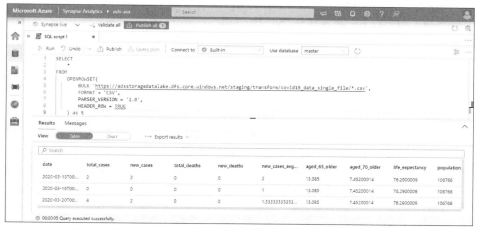

FIGURE 3-39 The OPENROWSET command output

Because OPENROWSET returns a full record set (speaking in terms of the SQL engine), you can perform further operations on it, like joining it with other tables or applying complex transformations on it. For example, Listing 3-11 shows how to use the OPENJSON function to parse the content of a JSON file, returning it in a tabular format. In this case, using the WITH clause is mandatory to specify the output schema, because the engine cannot infer it. (Remember that JSON is not one of the supported formats.)

LISTING 3-11 Using the OPENROWSET command with complex transformations

```
SELECT
  t.*
FROM
  OPENROWSET(
    BULK 'https://adsstoragedatalake.dfs.core.windows.net/globaldata/Daily/*',
    FORMAT = 'CSV',
    FIELDQUOTE = '0x0b',
    FIELDTERMINATOR ='0x0b',
    ROWTERMINATOR = '0x0b'
)
  WITH (jsonContent varchar(MAX)) AS j
  CROSS APPLY OPENJSON (jsonContent) with (
    [date] datetime2, total_cases int,
    new_cases int, total_deaths int,
    new_deaths int, total_cases_per_million decimal(13, 5),
    new_cases_per_million decimal(13, 5), total_deaths_per_million decimal(13, 5),
    new_deaths_per_million decimal(13, 5), stringency_index decimal(13, 5),
    new_cases_smoothed decimal(13, 5), new_deaths_smoothed decimal(13, 5),
    new_cases_smoothed_per_million decimal(13, 5), Code nvarchar(10),
    Country nchar(3)) as t
```

Although creating SQL views in a dedicated SQL pool might be straightforward, it might not be so obvious when targeting serverless SQL pools. Still, this is definitely possible. In fact, you can create one or more databases to "host" your traditional SQL views and PolyBase structures. These databases contain no data at all — just metadata and DML objects — and they will be always available when performing queries against the serverless SQL pool endpoint. Listing 3-12 shows how to create a database and a view. The view contains the same code as in Listing 3-11.

LISTING 3-12   Encapsulating the OPENROWSET command inside a view

```
CREATE DATABASE ads
GO
USE ads
GO
CREATE VIEW dbo.vw_covid_data as
SELECT
  t.*
FROM
  OPENROWSET(
    BULK 'https://adsstoragedatalake.dfs.core.windows.net/globaldata/Daily/*',
    FORMAT = 'CSV',
    FIELDQUOTE = '0x0b',
    FIELDTERMINATOR ='0x0b',
    ROWTERMINATOR = '0x0b'
)
  WITH (jsonContent varchar(MAX)) AS j
  CROSS APPLY OPENJSON (jsonContent) with (
    [date] datetime2, total_cases int,
    new_cases int, total_deaths int,
    new_deaths int, total_cases_per_million decimal(13, 5),
    new_cases_per_million decimal(13, 5), total_deaths_per_million decimal(13, 5),
    new_deaths_per_million decimal(13, 5), stringency_index decimal(13, 5),
    new_cases_smoothed decimal(13, 5), new_deaths_smoothed decimal(13, 5),
    new_cases_smoothed_per_million decimal(13, 5), Code nvarchar(10),
    Country nchar(3)) as t
```

Figure 3-40 shows the resulting database and the view contained in it.

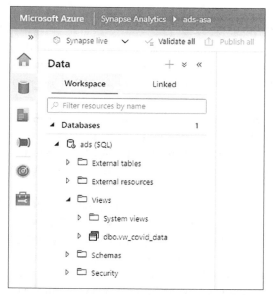

FIGURE 3-40 A database created on a serverless SQL pool

You can reach a serverless SQL endpoint from outside your Azure Synapse Analytics workspace through its public endpoint. This is in the following form:

```
<your_workspace_name>-ondemand.sql.azuresynapse.net
```

The `InitialCatalog` property of the connection string enables you to choose a specific database among the available ones in the pool. This way, you can, for example, issue a query from a Power BI dataset to read data stored directly in a blob storage, parsing it with logic encapsulated in a view in your Azure Synapse Analytics workspace.

In the same way, you can reach a dedicated SQL pool with its proper endpoint, which is in the following form:

```
<your_workspace_name>.sql.azuresynapse.net.
```

In this case, the `InitialCatalog` property of the connection string enables you to choose a specific pool among the available ones in the workspace.

Working with Spark pools

If you have a Microsoft SQL Server background, serverless and dedicated SQL pools should be very familiar. However, working with dedicated SQL pools requires you to approach data in a distributed manner. As you saw in Chapter 2, data in dedicated SQL pools does not reside in a single database (even though you see no big difference when accessing it). Rather, it is spread over 60 databases, called *distributions*. These are accessed by a scalable number of compute nodes, governed by a head node that serves external client requests and queries. When needed, the Data Movement Service (DMS) temporarily moves data (well, a copy of it) across distributions to be able to respond to a query from a client.

In this regard, Spark follows a very similar approach. Data is usually stored in a distributed file system, like HDFS or Azure storage, and is accessed by a scalable number of executor nodes under the directives of the driver node, which can be seen as the Spark counterpart of the head node in a dedicated SQL pool.

We discussed the principles behind Spark and how this engine works when we talked about Azure Databricks earlier in this chapter. We do not need to introduce these principles again, because they are still valid when working with Spark pools in Azure Synapse Analytics. In fact, the Spark engines behind Azure Databricks and Azure Synapse Analytics have common roots in the Apache Spark open-source project.

For projects that are based (almost) exclusively on Spark that require collaboration, with ETL pipelines authored in notebooks, the use of Delta Lake as a resilient file-based repository, ML workloads handled in SparkML, and so on, Azure Databricks probably remains the best choice. It offers a very mature environment tailored around the Spark engine. But when you prefer a mixed approach, having code-free ETL pipelines, a relational and scalable database engine as storage, a Spark workload for ML, and Power BI integration, Azure Synapse Analytics offers all of the above in the same workspace.

To author Spark code in Azure Synapse Analytics, you use Spark notebooks in the Develop Hub of Azure Synapse Studio. Of course, before you run Spark jobs, you must create a Spark pool in advance, like you did in the "Add an Apache Spark pool" practice exercise in Chapter 2. As in Azure Databricks, you have an already available SparkContext you can use to perform operations on your data. So, for example, you can simply write the PySpark code in Listing 3-13 in a notebook cell and run it to produce the same output from Listing 3-12.

LISTING 3-13 Read data in Spark with PySpark

```
df = spark.read.json('abfss://globaldata@adsstoragedatalake.dfs.core.windows.net/Daily/*')

display(df.limit(10))
```

A few things worth noticing:

- As you can see, Spark works natively with JSON files, and you do not need to perform complex transformations to be able to parse it.

- Because the storage account is linked to the Azure Synapse Analytics workspace, you just need to indicate the path you want to access. The engine takes care of passing the credentials for authentication and authorization.

- The data frame is created using the spark.read.json method. In that code, spark is shorthand for SparkContext.

The query in Listing 3-13 produces the following output, displayed in Figure 3-41.

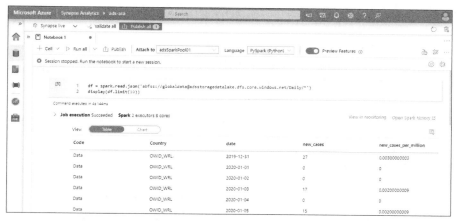

FIGURE 3-41 Read data in Spark with PySpark

Listing 3-14 performs the same operation, this time with C# and .NET Spark. You just need to either prefix your cell with the magic command %%csharp to switch languages in that particular cell or change the default language of the notebook with the drop-down selector along the top.

LISTING 3-14 Read data in Spark with C#

```
%%csharp
DataFrame df = spark.Read().Json("abfss://globaldata@adsstoragedatalake.dfs.core.windows.
net/Daily/*");
Display(df);
```

Reading and writing data to a dedicated SQL pool in the same workspace is very easy thanks to credential pass-through, which enables such scenarios with little or no configuration. To perform such operations, you can use the Azure Synapse Analytics connector, which is currently available only for Scala. The method to use is synapsesql, which accepts the target table name and type as parameters. The type could be INTERNAL (for traditional tables) or EXTERNAL (for external tables). Under the hood, it leverages the PolyBase component to move data as quickly as possible. Listing 3-15 shows the synapsesql command used to create a traditional table in a dedicated SQL pool.

LISTING 3-15 Write data to a dedicated SQL pool in Scala

```
%%spark
val scala_df = spark.read.json("abfss://globaldata@adsstoragedatalake.dfs.core.windows.net/
Daily/*")
scala_df.write.synapsesql("adsSQLpool01.dbo.covid_data", Constants.INTERNAL)
```

If you are authoring a PySpark notebook, you can share a data frame between Python and Scala contexts as follows:

1. Using the createOrReplaceTempView method, create an unmanaged Spark table over the source data frame, which is visible from the SparkSQL context. This step has to be done in a PySpark cell.

2. Using the SparkSQL context, create the target data frame. This step has to be done in a Scala cell.

Listing 3-16 and Listing 3-17 show the aforementioned approach. Because they must be run in different cells of a PySpark notebook, we separated the listings. Note the %%spark magic command in Listing 3-17, which indicates to Spark that the code is in Scala language.

LISTING 3-16 Create an unmanaged table in Spark with PySpark

```
%%pyspark
df = spark.read.json('abfss://globaldata@adsstoragedatalake.dfs.core.windows.net/Daily/*')
df.createOrReplaceTempView("vw_df")
```

LISTING 3-17 Create a Scala data frame out of the original PySpark data frame and write to a dedicated SQL pool

```
%%spark
val scala_df = spark.sqlContext.sql("select * from vw_df")
scala_df.write.synapsesql("adsSQLpool01.dbo.covid_data", Constants.INTERNAL)
```

> **NEED MORE REVIEW?** There are some limitations at the time of this writing — for example, the destination table must not exist. Read more here: *https://docs.microsoft.com/en-us/azure/synapse-analytics/spark/synapse-spark-sql-pool-import-export*.

Figure 3-42 shows that the target table has been created inside the adsSQLpool01 database.

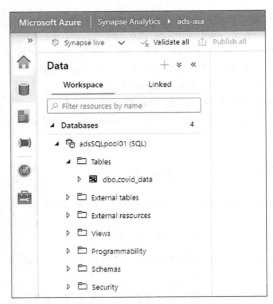

FIGURE 3-42 A table created with Spark on a dedicated SQL pool

If you prefer to maintain a clear separation between the Spark and SQL engines and, at the same time, have more flexibility on the database side, an alternative approach is to transform data in Spark, save the output in the data lake, and use a copy activity in a visual pipeline to ingest data in SQL. This way, you can do some setup on the target database (for example, dropping or truncating the target table if it already exists), execute the notebook that performs the needed transformation, and finally load the data into the target table — all orchestrated in an organic workflow.

Streaming data

A *data stream* is a continuous flow of information, where *continuous* does not necessarily mean regular or constant. A single chunk of raw information sent is an *event* (or *message*), and its size is rarely more than a few kilobytes. With some exceptions, the order of events does matter, so stream engines must implement some sort of reconciliation system to handle possible delays in the delivery. Figure 3-43 provides an overview of a stream pipeline and some of the technology involved.

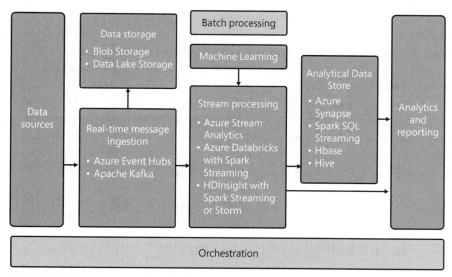

FIGURE 3-43 Stream-processing pipeline

To simplify and schematize a data-stream flow, you can identify the following main phases:

- **Production** Data is produced by various sources. These usually include one or more types of devices, such as (but not limited to) mobile phones, sensors, flight recorders, wind turbines, production lines, wearables, GPS applications, cameras, and software applications. These are the *producers*.

- **Acquisition** Produced data is pushed to one or more endpoints, where a stream transport and/or processing engine is listening for incoming data events. These events are made available to downstream clients. These are the *consumers*. The acquisition phase and consumers' queries are often referred to as *ingress* and *egress phases*, respectively. Hundreds, thousands, or even millions of producers can send data simultaneously and with very high frequency. So, having a low-latency and scalable engine (such as Azure Event Hubs, with or without a Kafka interface, or Apache Kafka) listening is mandatory on this side. Events are usually kept for a configurable period of time and left at the disposal of the consumers — not necessarily one by one but in small batches.

- **Aggregation and transformation** Once acquired, data can be aggregated or transformed. Aggregation is usually performed over time, grouping events by windows. Tumbling, hopping, sliding, and session windows are commonly used to identify specific rules for aggregation (more on this in a bit). Data can also undergo some transformation, such as filtering out unwanted values, enriching it by joining it with static datasets or other streams, or passing it to an ML service to be scored or as a target of prediction. Aggregated data is then stored or sent to a real-time-capable dashboard tool, like Microsoft Power BI, to provide users with a constantly fresh and insightful view of information flowing in.

- **Storage** Acquired data — raw, aggregated, or both — is stored for further analysis. Storage types may vary and depend on whether aggregation has been performed. Raw data is usually sent to data lake folders in compressed format or high-ingestion-throughput NoSQL database services like Azure Cosmos DB. Aggregated data, on the other hand, is typically stored in data lake folders in compressed format or relational database services such as Azure SQL Database or Azure Synapse Analytics.

Figure 3-44 shows a typical use case for streaming: road vehicle trips analysis.

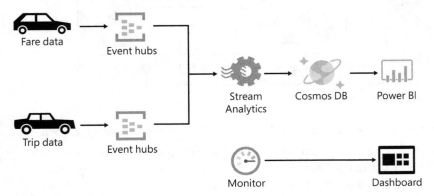

FIGURE 3-44 Stream-processing overview

NOTE Message-ingestion engines do not usually keep events forever. Rather, they delete them after a configurable retention period. Azure Event Hubs have a feature called *Event Hub Capture* that you can use to offload incoming events to cloud storage as soon as they arrive, as well as pass them down your stream pipeline. This feature is useful when you need to run a batch-processing job afterward or you want to keep events for historical reasons as it prevents you from having to build an offload pipeline yourself. Events are serialized in Apache Avro, a compact and binary serialization format that retains schema information.

Stream-transport and processing engines

Stream-transport and processing engines are complex pieces of software. They usually operate 24/7; hence, resiliency against failure is a key factor. Moreover, they must scale quickly as soon as the volume of incoming data increases, since losing events is not an option.

From a high-level point of view, these usually come as a classical cluster, with a driver node for coordination and a variable number of executor nodes that do the physical work. If you choose to use platform as a service (PaaS) services, this architecture will likely be transparent to you.

NOTE One of the most important metrics to check the health of your pipeline is the input rate versus the processing rate (InputRate/ProcessingRate) coefficient. It shows how effective your pipeline is in ingesting and transforming data as soon as it arrives. If you have a high value for this ratio, it means one of two things:

- Your processing engine is under too much pressure and needs to scale.

- You are doing too many transformations to your incoming messages or the messages are too complex.

Data streams share some concepts that are important to understand, considering that they are unique to these workloads and not so common in other processes:

- **Watermarks**   In such complex systems with so many actors communicating at the same time, it is almost impossible to have no failure or disruption. Be it a network error, a producer losing a signal, or a cluster node crashing, events must be delivered and kept in order. Watermarking is a technique used by stream-processing engines to deal with late-arriving events. A watermark is basically a threshold; when a message is sent, if the difference between the arrival time and the production time is above this threshold, the message is discarded from the output and is not sent to the consumer.

- **Consumer groups** These allow consumers to maintain dedicated and isolated views over the data stream. The source stream is unique, whereas each group can read data at its own pace and starting from its own offset. For example, you might have a real-time dashboard that reads data every five seconds and an hourly job that performs historical aggregations over ingress events. Both are reading the same stream, but the former will read events more frequently than the latter.
- **Time window aggregations** These define how events should be grouped to produce the desired output. They are explored in detail in the "Select the appropriate built-in functions" section later in this chapter.

Every engine implements these concepts in their own way. This chapter focuses on a very powerful engine that is part of the Azure offering: Azure Stream Analytics.

Implement event processing using Stream Analytics

Azure Stream Analytics is a fully managed platform as a service (PaaS) offering on Azure. It can run in the cloud as well as on the IoT Edge runtime and Azure Stack. This allows for ultra-low latency analytics because it can operate very close to many producers of events.

Azure Stream Analytics can process millions of events every second, and its ability to scale up and scale out allows for good performance even with non-uniform flows of events. Also, its elasticity enables you to keep your costs as low as possible, consuming just those resources needed.

Typical scenarios include the following:

- Analyzing real-time telemetry streams from IoT devices
- Web logs and clickstream analytics
- Geospatial analytics for fleet management
- Remote monitoring and predictive maintenance of high value assets
- Real-time analytics over streams of data for anomaly detection

With respect to reliability, Azure Stream Analytics guarantees:

- **Exactly-once processing** Given a set of inputs, the system will always return the same results. This feature is very important for repeatability, especially in the case of service downtime or disruption.
- **At-least-once delivery** All outputs from exactly-once processing are delivered to the output, but duplicate results may occur. However, with output such as CosmosDB or SQL Server, exactly-once delivery may be achieved. For the former, no action is needed because it uses upsert operations. For the latter, you need a natural key for outputted streaming events and a unique constraint (or a primary key) on the natural key in the destination table.

> **NEED MORE REVIEW?** You can read about event delivery guarantees by Azure Stream Analytics here: *https://docs.microsoft.com/en-us/stream-analytics-query/event-delivery-guarantees-azure-stream-analytics.*

An Azure Stream Analytics job consists of the following:

- **One or more inputs** Supported sources of events include Azure Event Hubs (with or without Kafka interface), Azure IoT Hub, and Azure Blob storage.

- **A query** A SQL-like language is used to perform analysis and aggregation over incoming events. Along with classical aggregation and filter functions, you can join streaming events with static datasets (hosted on an Azure SQL Database or Azure Blob storage) to enrich the data or you can integrate Azure Machine Learning predictions. Events can be aggregated within time windows using specific statements, and additional extensibility is achievable through user-defined functions (UDFs) written in C# or JavaScript.

- **One or more outputs** The result of the query performed over the data is sent to listening consumers. Typical sinks include the following:

  - Azure Functions or Azure Service Bus to trigger custom workflows downstream
  - Power BI dashboards to perform real-time monitoring
  - Data stores (such as Azure Blob storage, ADLS, and Azure Synapse Analytics) to make data available to further analysis or batch processes

Provision an Azure Stream Analytics job

To provision an Azure Stream Analytics job from the Azure Portal, follow these steps:

1. Open *https://portal.azure.com* and use your credentials to log in.

2. Type **Stream Analytics job** in the search box and select the corresponding entry that appears in the drop-down list.

3. In the Stream Analytics Job page, shown in Figure 3-45, click the **Create** button.

FIGURE 3-45 The Azure Stream Analytics Job page

4. In the **New Stream Analytics Job** page, configure the following settings:
   - **Job Name**   Type a meaningful name for the job.
   - **Subscription**   Enter the subscription in which to create the job.
   - **Resource Group**   This is the resource group in which the job will be created.

   > **NOTE**   If the resource group does not yet exist, you can create it from here.

   - **Location**   Enter the Azure region in which to create the job.
   - **Hosting Environment**   Specify whether the job will be deployed in Azure or on an on-premises IoT Gateway Edge device.
   - **Streaming Units**   Enter the number of computation resources that will be available to process the query. This can be a value between 1 and 192.
5. Click the **Create** button.

   The Azure Portal generates the necessary ARM template, sends it to be deployed, and displays a page with a *Your deployment is in progress* message, informing you that it is currently in the creation phase.

Author a job

After you provision your Azure Stream Analytics job, you can start authoring the job. You begin from the job's Overview page. (See Figure 3-46.)

FIGURE 3-46   Azure Stream Analytics Overview page

For the moment, the job is actually paused. This is because it is missing its core parts: inputs, outputs, and a query. These parts together form the job topology. The steps needed to define inputs and outputs are described in the next section.

When it processes incoming events, an Azure Stream Analytics job consumes Streaming Units (SU). You can scale up and scale down the number of SUs available for the job clicking the Scale button in the left toolbar, in the Configure category. On the Scale page, shown in Figure 3-47, a slider enables you to move the number of streaming units between 1 and 192. (Note that there is no option for autoscaling.)

FIGURE 3-47 Scaling an Azure Stream Analytics job

Configure input and output

Inputs of an Azure Stream Analytics job are divided into two categories:

- **Data stream** An unbounded stream of events. At least one data stream must be present in a job. Supported sources are as follows:
    - **Event Hubs** Used to collect event streams from multiple devices and services, such as social media activity feeds, stock trade information, and data from sensors.
    - **IoT Hubs** Used to collect data from connected devices in Internet of Things (IoT) scenarios.
    - **Azure Blob storage** Used to ingest bulk data as a stream, such as log files from a server or an application.
    - **ADLS Gen2** Used to ingest bulk data as a stream, such as log files from a server or an application.
- **Reference data** This input is static or changes very slowly over time. Reference data is usually joined to a data stream to enrich incoming events. For example, consider a data stream with a sensor ID and reference data pulled from a database with detailed information about the sensor. Supported sources are:
    - Azure Blob storage

- ADLS Gen2
- Azure SQL Database

> **NEED MORE REVIEW?**   For a complete overview of supported data stream inputs, see *https://docs.microsoft.com/en-us/azure/stream-analytics/stream-analytics-define-inputs*. To get a complete overview of supported reference data inputs, see *https://docs.microsoft.com/en-us/azure/stream-analytics/stream-analytics-use-reference-data*.

Azure Stream Analytics has native support for processing events in the following formats:

- CSV
- JSON
- Avro

Both JSON and Avro may contain complex types, like nested objects and arrays.

> **NOTE**   Avro's built-in deserializer does not support the **Map** type (which is basically a key-value pair collection). So, Avro files produced by Azure Event Hubs Capture cannot be read by Azure Stream Analytics unless you use a custom .NET deserializer.

Add an input to a job

To add an input to your job, you click the **Inputs** button in the **Job Topology** category in the left toolbar of the job resource page in Azure Portal. The Inputs page opens. This page has two buttons along the top:

- **Add Stream Input**   There are three kinds of stream inputs:
    - Event Hub stream input
    - IoT Hub stream input
    - Blob storage/ADLS Gen 2 stream input
- **Add Reference Input**   There are two types of reference inputs:
    - Blob storage/ADLS Gen 2 reference input
    - SQL Database reference input

Whichever input you choose, a tab will appear on the right, where you can configure and create your input.

EVENT HUB STREAM INPUTS

An Event Hub stream input requires you to configure the following settings:

- **Input Alias**   The name of the input. This will be used to reference it in a query.
- **Event Hub**   Choose one from an accessible subscription, enter its endpoint manually, or create a new one.

- **Consumer Group**   The consumer group of the Event Hub that will host the reader process. Although a consumer group allows for five readers at a time, it is highly recommended that you have a dedicated consumer group for each job. You can pick an existing one or create a new one.

- **Authentication Mode**   The connection string to the Event Hub. Alternatively, you can use your service's Managed Identity (in public preview at the time of this writing) instead of the connection string.

- **Partition Key**   If your input is partitioned by a property, you can specify the property name. This will improve the performance of your query when using group by or partition by clauses.

- **Event Serialization Format**   The serialization format used by your input data stream. Your choices are JSON, Avro, CSV, and Other. If you choose Other, you will need a custom deserializer written in .NET. If you choose JSON or CSV instead, you will need to specify the encoding and compression used by input events. In addition, for CSV, you will have to provide a column delimiter.

IOT HUB STREAM INPUTS

An IoT Hub stream input requires you to configure the following settings:

- **Input Alias**   The name of the input. This will be used to reference it in a query.

- **IoT Hub**   Choose one from an accessible subscription or enter its endpoint manually.

- **Consumer Group**   The consumer group of the IoT Hub that will host the reader process. Although a consumer group allows for five readers at a time, it is highly recommended that you have a dedicated consumer group for each job. You can pick an existing one or create a new one.

- **Shared Access Policy**   The shared access policy name and key for connecting to the IoT Hub.

- **Endpoint**   Choose Messaging to listen to messages from devices to the cloud or Operations Monitoring to listen to devices' telemetry and metadata.

- **Partition Key**   If your input is partitioned by a property, you can specify the property name. This will improve the performance of your query when using group by or partition by clauses.

- **Event Serialization Format**   The serialization format used by your input data stream. Your choices are JSON, Avro, CSV, and Other. If you choose Other, you will need a custom deserializer written in .NET. If you choose JSON or CSV instead, you will need to specify the encoding and compression used by input events. In addition, for CSV, you will have to provide a column delimiter.

BLOB STORAGE/ADLS GEN2 STREAM INPUTS

A Blob storage/ADLS Gen2 stream input requires the following information:

- **Input Alias**   The name of the input. This will be used to reference it in a query.

- **Storage Account** Choose one from an accessible subscription, enter its name manually, or create a new one.

- **Container** The name of the container to monitor for changes. You can select an existing one or create a new one.

- **Authentication Mode** The connection string to the storage account. Alternatively, you can use your service's Managed Identity (in public preview at the time of this writing) instead of the connection string.

- **Path Pattern** The file path used to locate your blobs within the specified container. You can make the path dynamic, combining one or more instances of the **{date},** **{time},** and **{partition}** variables. For example, you can enter values like **telemetry/ devices/{partition}/{date}/{time}**.

- **Date Format** The format used to parse {date} variable content.

- **Time Format** The format used to parse {time} variable content.

- **Partition Key** If your input is partitioned by a property, you can specify the property name. This will improve the performance of your query when using group by or partition by clauses.

- **Count of Input Partitions** The upper limit for the {partition} variable when used inside the path pattern. When this variable is present in the pattern, the job will look for blobs in a partition range between 0 and the number of input partitions –1.

- **Event Serialization Format** The serialization format used by your input data stream. Your choices are JSON, Avro, CSV, and Other. If you choose Other, you will need a custom deserializer written in .NET. If you choose JSON or CSV instead, you will need to specify the encoding and compression used by input events. In addition, for CSV, you will have to provide a column delimiter.

BLOB STORAGE/ADLS GEN2 REFERENCE INPUTS

A Blob storage/ADLS Gen2 reference input requires you to configure the following settings:

- **Input Alias** The name of the input. This will be used to reference it in a query.

- **Storage Account** Choose one from an accessible subscription, enter its name manually, or create a new one.

- **Container** The name of the container that contains the reference files. You can select an existing one or create a new one.

- **Authentication Mode** The connection string to the storage account. Alternatively, you can use your service's Managed Identity (in public preview at the time of this writing) instead of the connection string.

- **Path Pattern** The file path used to locate your reference file within the specified container. You can make the path dynamic, combining one or more instances of the **{date}** and **{time}** variables. For example, you can enter values like **telemetry/devices/{date}/**

devices_list.csv. This is useful when you have a reference input that is updated periodically and you want to use a specific snapshot of it in your job for consistency.

- **Date Format**   The format used to parse {date} variable content.
- **Time Format**   The format used to parse {time} variable content.
- **Event Serialization Format**   The serialization format used by your reference file. Your choices are JSON, Avro, CSV, and Other. If you choose Other, you will need a custom deserializer written in .NET. If you choose JSON or CSV instead, you will need to specify the encoding used by the reference file. In addition, for CSV, you will have to provide a column delimiter.

SQL DATABASE REFERENCE INPUTS

A SQL Database reference input issues a query to the specified database and stores the result in a storage account. You can retrieve the data as soon as the job starts or keep polling the source at specified time intervals. This input requires you to configure the following settings:

- **Input Alias**   The name of the input. This will be used to reference it in a query.
- **Server Name**   The SQL Server that contains the database.
- **Database**   The database that contains the reference data.
- **Storage Account**   The storage account that will contain the reference data outputted by the query.
- **Username and Password**   The credentials to access the SQL Server.
- **Snapshot Query**   The T-SQL query to extract the reference data. If the source is a temporal table, you can use the following syntax to retrieve a specific snapshot of the records:

    `FOR SYSTEM_TIME AS OF @snapshotTime`.

- **Refresh Periodically**   Whether the query has to be issued at specific intervals or just one time as the job starts. The minimum interval is one minute, and the maximum interval is 365 days. You can also specify a query to fetch incremental changes from the database for the reference data. A temporal table as a source is recommended in this scenario.

Outputs

A query in an Azure Stream Analytics job combines and processes one or more inputs to produce one or more outputs. There are many supported outputs, and all of them require you to provide an alias; connection and authentication information; the destination table, file, topic, or queue; and the partition schema to be used (if supported). Table 3-1 displays all the available outputs, whether they support partitioning, and the enabled authentication modes.

TABLE 3-1 Outputs of an Azure Stream Analytics job

Output Type	Partitioning	Security
ADLS Gen 1	Yes	Azure Active Directory user Managed Identity
Azure SQL Database	Yes (optional)	Access key Managed Identity (preview)
Azure Synapse Analytics	Yes	Access key Managed Identity (preview)
Blob storage and ADLS Gen 2	Yes	Access key Managed Identity (preview)
Azure Event Hubs	Yes (need to set the partition key column in output configuration)	Access key Managed Identity (preview)
Power BI	No	Azure Active Directory user Managed Identity
Azure Table storage	Yes	Account key
Azure Service Bus queues	Yes	Access key
Azure Service Bus topics	Yes	Access key
Azure Cosmos DB	Yes	Access key
Azure Functions	Yes	Access key

> **NEED MORE REVIEW?** For detailed information about supported outputs and their configuration, read here: *https://docs.microsoft.com/en-us/azure/stream-analytics/ stream-analytics-define-outputs.*

As you can see in Table 3-1, Azure Stream Analytics supports partitions for all outputs except for Power BI. The number of partitions in your output data determines how many writers will be used when writing to the sink data store. In addition, you can tune the number of partitions using the INTO <partition count> clause in your query.

If your output adapter is not partitioned and one input partition lacks data, the writer process will wait for incoming data depending on the late arrival policy before merging all inputs in a single stream and sending it to the sink. This can be the source of bottlenecks in your pipeline.

> **NOTE** Event ordering and time handling are important concepts in every stream-processing engine. Read more here: *https://docs.microsoft.com/en-us/azure/stream-analytics/ stream-analytics-time-handling.*

Although all outputs support batching, only some of them support an explicit batch size. In fact, Azure Stream Analytics uses variable-size batches to process events and write to outputs. The higher the rate of incoming/outgoing events, the larger the batches. When the egress rate is low, small batches are preferred to keep latency low.

To add an output to your job, you click the **Outputs** button in **Job Topology** category in the left toolbar of the job resource page in Azure Portal. The Outputs page opens. Like the Inputs page, the Outputs page has two buttons along the top: Add Stream Output and Add Reference Output.

Whichever output one you choose, a tab will appear on the right, where you can configure and create your output. For the most part, output configuration is similar to input configuration, so we do not cover it in detail here. However, you can refer to the external links provided throughout this section for an in-depth overview of each output option.

Select the appropriate built-in functions

After you set up the inputs and outputs for your job, your last step is to create a query that processes incoming events to produce the desired output(s).

Queries in Azure Stream Analytics are expressed in a SQL-like language. They can contain simple pass-through logic that just moves source events to output data stores, or they can perform more complex transformations like pattern matching and temporal analysis to calculate aggregates over various time windows.

Data from multiple inputs can be joined to combine streaming events, and you can perform lookups against reference data to enrich event values. A query is not limited to just one output because it can write events to multiple outputs at once.

Listing 3-18 shows a sample pass-through query in Azure Stream Analytics. As you can see, there is no particular logic or transformation applied to the input data. The INTO clause is used to specify the output that will receive the incoming events.

LISTING 3-18 Pass-through query

```
SELECT
    *
INTO Output
FROM Input
```

Listing 3-19 shows a query that aggregates data over time. In this example, the query counts all tweets received as input every 10 seconds, aggregating them by country.

LISTING 3-19 Aggregation over time

```
SELECT
    Country,
    COUNT(*) AS TweetsCount
FROM
    Input TIMESTAMP BY Time
GROUP BY
    Country,
    TumblingWindow(second, 10)
```

Notice that Listing 3-19 defines the 10-second window using a function that is not part of the familiar T-SQL syntax: TumblingWindow. This is a built-in function that is specific to the Azure Stream Analytics query language. Built-in functions can be grouped in the following categories:

- **Aggregate functions**   These operate on collections of values, returning a single summary value. Examples of these functions are AVG, COUNT, MAX, and SUM.

- **Analytic functions**   These return a value based on a defined constraint. Examples of these functions are ISFIRST, LAG, and LEAD.

- **Array functions** These return information from an array. Examples of these functions are GetArrayElement, GetArrayElements, and GetArrayLength.

- **GeoSpatial functions** These operate on spatial data to perform real-time geospatial analytics. Examples of these functions are CreatePolygon, ST_DISTANCE, and ST_INTERSECTS.

- **Input Metadata functions** These return input-specific properties — for example, a deterministic unique EventId for each incoming event. The only function in this category is GetMetadataPropertyValue.

- **Record functions** These return record-specific properties or values. They are very useful when dealing with complex input formats, like those that contain nested objects. Examples of these functions are GetRecordProperties and GetRecordPropertyValue.

- **Windowing functions** These functions perform operations on events within a time window. Examples of these functions are TumblingWindow, HoppingWindow, SlidingWindow, and SessionWindow.

- **Scalar functions** These operate on a single value and return a single value. Scalar functions can be further divided in four subcategories:

  - **Conversion functions** These can be used to cast data into different formats. Examples of these functions are CAST and TRY_CAST.

  - **Date and Time functions** These operate on date/time formats. Examples of these functions are DATEPART, YEAR, DATEDIFF, and DATEADD.

  - **Mathematical functions** These perform calculations on input values passed as arguments and return a numeric value. Examples of these functions are ABS, FLOOR, SQRT, and ROUND.

  - **String functions** These operate on strings. Examples of these functions are LEN, CONCAT, LOWER, UPPER, SUBSTRING, and REGEXMATCH.

> **NEED MORE REVIEW?** To learn more about built-in functions in Azure Stream Analytics, see *https://docs.microsoft.com/en-us/stream-analytics-query/built-in-functions-azure-stream-analytics.*

Most of these functions adhere to the same rules as the T-SQL language. For example, you can use aggregate functions only in the select list of a SELECT statement or in a HAVING clause. Scalar functions, on the other hand, can be used wherever an expression is allowed.

In a streaming scenario, it is very common to perform set-based operations over a subset of events that fall within a specific time window. For this reason, it is worthwhile to explore time window aggregations in more detail.

To better understand the logic behind time window aggregations (see Figure 3-48), let us take a closer look at four types of them that you may need to apply in your query, all available in Azure Stream Analytics through the corresponding built-in windowing function:

- Tumbling window
- Hopping window

- Sliding window
- Session window

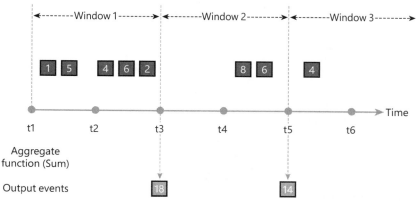

FIGURE 3-48   A stream window aggregation

Tumbling window

A tumbling window is a fixed-size segment of time that repeats and does not overlap with its preceding segment. (See Figure 3-49.) Events can belong to just one window, and if no events occur in a specific time window, the window will be empty. One of the most typical uses of tumbling windows is to aggregate data for reporting purposes — for example, counting the number of financial transactions that occurred in the last hour and storing the result.

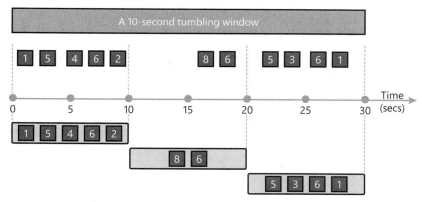

FIGURE 3-49   Tumbling window example

Hopping window

A hopping window (see Figure 3-50) has two parameters:

- **Hop Size**   Indicates how much the window must hop forward in time
- **Window Size**   Indicates how many seconds it has to go back in time to collect events

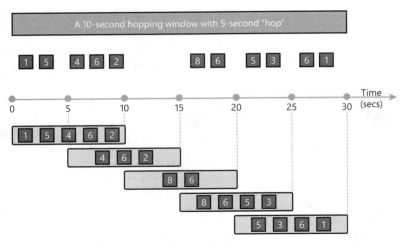

FIGURE 3-50 Hopping window example

Windows can overlap, they can be empty, and events can belong to multiple windows.

If it makes it simpler, think of a hopping window as a tumbling window that can overlap. Indeed, if the Hop Size and Window Size settings are the same, a hopping window will behave exactly like a tumbling window. A typical use of a hopping window is to compute moving averages over incoming data.

Sliding window

A sliding window moves forward in time in fixed intervals and looks back by a specific size (see Figure 3-51). However, it does not produce output if no new events occur. Windows can overlap, they cannot be empty, and events can belong to multiple windows.

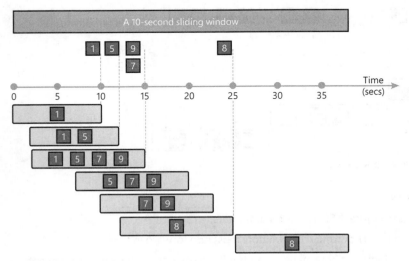

FIGURE 3-51 Sliding window example

As with hopping windows, sliding windows are often used to compute moving averages. The difference is that while hopping windows are computed at fixed intervals, sliding windows adjust their frequency with the density of incoming messages, producing more accurate results when events are very close.

Session window

A session window has three parameters:

- Timeout
- Maximum Duration
- Partition Key

The first two are mandatory and the last one is optional.

Figure 3-52 shows a session window with a five-minute timeout and a 20-minute maximum duration. It works as follows:

1. When the first event arrives, a new window is created.

2. A second event comes before five minutes have passed. Because the window is still waiting for new messages, it is extended, and the timeout is reset.

3. The engine waits another five minutes. Because no new messages appear, the window is closed. The sum aggregate value for this particular window would be 6 (1 +5).

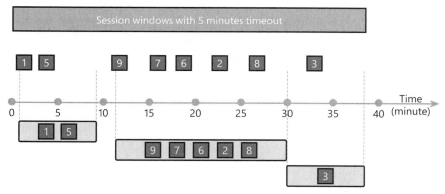

FIGURE 3-52   Session window example

Generally speaking, when an event occurs, a new window is created. If no events arrive within the specified timeout, the window is closed. Otherwise, it extends and keeps waiting for other events to flow in. If it reaches its Timeout value, or it extends to its maximum duration, the window is closed. In Figure 3-52, the maximum duration is never reached, because a timeout always occurs before it.

Here are a few other important points:

- If a partition key has been specified, it acts as a boundary; the same process applies to every partition value, without interfering with the others.

- Windows cannot overlap, they cannot be empty, and events cannot belong to multiple windows.
- Session windows are useful when you want to analyze events together that may be related, like user interactions within a website or an app (visited pages, clicked banners, and so on).

> **NEED MORE REVIEW?**   Read more about windowing functions in Azure Stream Analytics here: *https://docs.microsoft.com/en-us/stream-analytics-query/ windowing-azure-stream-analytics.*

Edit the query for an Azure Stream Analytics job

To edit the query for your Azure Stream Analytics job, click the **Query** button in the **Job Topology** category in the left toolbar of the job resource page in the Azure Portal. The Query page opens. (See Figure 3-53.) You can edit, test, and save your query in the main pane. On the left, your inputs and outputs are listed. At the bottom is a preview of the input events (you can also upload a sample input to emulate a stream) and of the output results that will be produced. Finally, along the top is a Query Language Docs menu, which you can click to select from a list of relevant Microsoft documents; an Open in Visual Studio button, to edit your job in Visual Studio or Visual Studio Code; and a UserVoice button, which you can click to share your feedback about Azure Stream Analytics.

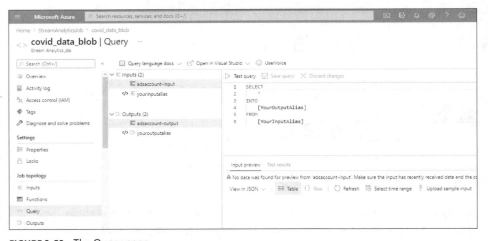

FIGURE 3-53   The Query page

> **NOTE**   To be able to develop an Azure Stream Analytics job in Visual Studio 2013 or higher, you first need to install Azure Data Lake and Stream Analytics Tools. You can download these here: *https://docs.microsoft.com/en-us/azure/stream-analytics/ stream-analytics-tools-for-visual-studio-install.*

PRACTICE Create an Azure Stream Analytics job

This practice exercise shows you how to create a simple Azure Stream Analytics job. The job will monitor a blob container for uploads of COVID data files from different countries. (These will be your input events.) It will then perform some summarizations over the input data. First, you will provision an Azure Stream Analytics job. Then you will set up your inputs, outputs, and query.

You already uploaded the files you'll use here to a storage account in Exercise 2.3 and Exercise 2.4 in Chapter 2. You can use the same storage account for this practice exercise, copying a single file (or batches of them) at a time to the monitored blob container to emulate uploads from remote locations. Alternatively, you can use the tool of your choice (for example, Azure Storage Explorer, the Azure web portal, PowerShell, or AzCopy) to upload the COVID data files from the Data folder in the companion content to the monitored blob container.

1. Open *https://portal.azure.com* and use your credentials to log in.

2. Click the **+Create a Resource** button.

3. Type **Stream Analytics job** in the search box and select the corresponding entry that appears in the drop-down list.

4. In the Stream Analytics Job page, click the **Create** button.

5. In the **New Stream Analytics Job** page, configure the following settings:

   - **Job Name**   Type **covid_data**.
   - **Subscription**   Select the subscription you have used in previous practice exercises.
   - **Resource Group**   Select the resource group you have used in previous practice exercises.
   - **Location**   Choose the same region you used in previous practice exercises (for example, **West Europe**).
   - **Hosting Environment**   Select **Cloud**.
   - **Streaming Units**   Type **1**.
   - **Secure All Private Data Assets Needed by This Job in My Storage Account**   Leave this unchecked.

6. Click the **Create** button.

7. Wait for the job to be created. Then navigate to the job's **Overview** page and click the **Input** button in the **Job Topology** section of the left toolbar.

8. Click the **Add Stream Input** button and select **Blob Storage/ADLS Gen2** in the list that opens.

9. In the **Blob Storage/ADLS Gen2** tab, configure the following settings:

   - **Input Alias**   Type **BlobInput**.
   - **Storage Account**   Use the same storage account you've used in previous practice exercises.

- **Container**   Use the same container you've used in previous practice exercises or create a new one here.
- **Authentication Mode**   Choose **Connection String**.
- **Path Pattern**   Type **InputData**. The job will monitor this folder for new files.
- **Partition Key**   Leave this box empty.
- **Event Serialization Format**   Leave this set to **JSON**.
- **Encoding**   Leave this set to **UTF-8**.
- **Event Compression Type**   Leave this set to **None**.

10. Click **Save** and wait for the input to be validated and created.

11. Click the **Add Reference Input** button and select **Blob Storage/ADLS Gen2** in the list that opens.

12. In the **Blob Storage/ADLS Gen2** tab, configure the following settings:
    - **Input Alias**   Type **BlobReferenceInput**.
    - **Storage Account**   Use the same storage account you've used in previous practice exercises.
    - **Container**   Use the same container you've used in previous practice exercises or create a new one here.
    - **Path Pattern**   Type **ReferenceData/covid-data_Countries.json**.
    - **Date Format**   Leave this set to **YYYY/MM/DD**.
    - **Time Format**   Leave this set to **HH**.
    - **Event Serialization Format**   Leave this set to **JSON**.
    - **Encoding**   Leave this set to **UTF-8**.
    - **Authentication Mode**   Select **Connection String**.

13. Click **Save** and wait for the input to be validated and created.

14. Click the **Output** button in the **Job Topology** section of the left toolbar in the job's **Overview** page.

15. Click the **Add** button and select **Blob Storage/ADLS Gen2** in the list that opens.

16. In the **Blob Storage/ADLS Gen2** tab, configure the following settings:
    - **Output Alias**   Type **BlobOutput**.
    - **Storage Account**   Use the same storage account you've used in previous practice exercises.
    - **Container**   Use the same container you've used in previous practice exercises or create a new one here.
    - **Path Pattern**   Type **OutputData/{date}/{time}**. This is the folder where the job will store output.
    - **Date Format**   Leave this set to **YYYY/MM/DD**.

- **Time Format**   Leave this set to **HH**.
- **Event Serialization Format**   Leave this set to **JSON**.
- **Encoding**   Leave this set to **UTF-8**.
- **Format**   Choose **Array**.
- **Minimum Rows**   Leave this field empty.
- **Hours**   Leave this field empty.
- **Minutes**   Leave this field empty.
- **Authentication Mode**   Select **Connection String**.

17. Click **Save** and wait for the output to be validated and created.

18. In the left Azure Portal toolbar, under **Job Topology**, click **Query**.

19. In the left toolbar of the **Query** page, under **Inputs**, click **BlobInput**.

20. In the bottom pane, in the **Input Preview** tab, click **Upload Sample Input**.

21. Locate and select the COVID daily data file named **covid-data_Data.ABW.json** in the companion content and click **OK**.

    The bottom pane will refresh to show a sample of the file contents in tabular format. The table contains columns like Date, total_cases, new_cases, and so on. You can also switch to Raw view to look at the JSON file content in its original format. This file will be used as a source for events when you test the query.

22. In the left toolbar of the **Query** page, under **Inputs**, click **BlobReferenceInput**.

23. In the bottom pane, in the **Input Preview** tab, click **Upload Sample Input**.

24. Locate and select the COVID demographic file named **covid-data_Countries.json** in the companion content and click **OK**.

    The bottom pane will refresh to show a sample of the file contents in tabular format. The table contains columns like Code, continent, location, and so on. You can also switch to Raw view to look at the JSON file content in its original format. This file will represent your reference data. It will be joined with the sample events when you test the query, enriching the incoming stream with country information.

25. Enter the query in Listing 3-20 in the text editor in the main pane of the Query page.

> **NOTE**   This query is also available from the covid_data.sql file, found in the ASA file in the companion content. If you prefer, you can copy it there and paste it into the text editor in the Query page.

LISTING 3-20   Aggregation over time

```
SELECT
  System.Timestamp() as OutTimes,
  ref.Continent,
  MAX(input.new_cases) as MaxNewCasesReported
```

```
INTO
  [BlobOutput]
FROM
  [BlobInput] as input
LEFT JOIN
  [BlobReferenceInput] as ref ON input.Country = ref.CountryCode
GROUP BY
  ref.Continent,
  HOPPINGWINDOW(minute, 3, 1)
```

26. Take a moment to look at the query. Try to guess what it does and what its output should be.

   To better understand the code, let's divide it into logical its operations:

   ■ The query reads events from the BlobInput stream input in the FROM clause. It then enriches it by joining it with the BlobReference reference input with a JOIN clause. Because you are dealing with an Azure Blob storage account input type, Azure Stream Analytics uses the blob creation time as the event arrival date. To modify this behavior, you can use the TIMESTAMP BY clause to specify a column or property to supply the time information for subsequent aggregations and computations.

   ■ The group by clause groups data by a column, named Continent (from the reference data file), and by a hopping window built-in function, HOPPINGWINDOW(minute, 3, 1). The hopping window is defined as a three-minute window that moves ahead one minute at a time. In other words, the engine will aggregate data every minute, considering only the events that arrived in the last three minutes. Because the input is an Azure Blob storage account, this translates to, "output will be produced only if new files have been placed in the folder in the last three-minute interval."

   ■ A traditional SELECT clause handles the projection part of the query. It outputs the columns named Continent (from the reference data input), the maximum number of new cases reported, and the OutTime field, computed as System.Timestamp(). This last field represents the upper boundary of the hopping window for each resulting group.

   ■ The INTO clause writes the results to the BlobOutput output.

27. Click **Save Query** to save the query.

28. Click **Test Query** to make a test run.

   The bottom panel switches to the Test Results tab. After a few seconds, a table containing three records is displayed: one for the current minute and two for the next two minutes. (This simulates subsequent outputs for the duration of the hopping window.) This table contains three columns: OutTimes, Continent, and MaxNewCasesReported.

   Because you have set up all the required parts and your query runs fine, you are almost ready to start the job. First, though, because you've uploaded just a sample of the reference data so far, you need to put the actual file in the correct folder so the job will be able to pick it up at runtime.

29. Open the **/Demography** folder and the **globaldata** container. Then locate the **covid-data_Countries.json** JSON file. (This file is also available in the companion content.)

30. Copy (or upload) the file to the **ReferenceData** folder in the container you have chosen as the BlobReference.

> **NOTE** If no ReferenceData folder exists, you can create one.

31. Click **Overview** in the left toolbar of the Azure Portal.

32. Click the **Start** button along the top of the page.

33. In the **Start Job** tab, leave the **Job Output Start Time** option set to **Now** and click **Start**.

You will be notified that a job is starting, and after about a minute, the job will start.

34. Click the **Monitoring** chart at the bottom of your resource's **Overview** page.

You will be redirected to the Metrics page, which displays a chart with three metrics: Input Events, Output Events, and Runtime Errors. After about a couple of minutes, you will see a spike of 211 events in the Input Events line. This is the number of reference events in the covid-data_Countries.json file you uploaded previously. No stream input events have been received at the moment.

Next you will upload files to the InputData folder to simulate an incoming stream.

35. Open the **/Daily** folder and the **globaldata** container. Then locate the **ABW** file. (This file is also available in the companion content.)

36. Copy (or upload) the file to the **InputData** folder in the container you have chosen as the BlobInput.

> **NOTE** If no InputData folder exists, you can create one.

37. Click the **Monitoring** chart at the bottom of your resource's **Overview** page to open the Metrics page.

After about a minute, you will see a spike of 179 events in the Input Events line. (If the chart does not refresh automatically, click **Refresh** at the top of the page). These are the events contained in the ABW file you just uploaded.

Wait a few more seconds, and you will see that the Output Events count increases by one. (You can hover the mouse pointer over the **Output Events** label below the chart to highlight that specific line.)

38. Open a new browser tab, log into the Azure portal, and go to the Overview page for the storage account you selected as **BlobOutput**.

39. Click **Azure Storage Explorer (Preview)** in the left panel. Then open the **OutputData** folder in the **BlobOutput** blob container and navigate to the innermost subfolder.

    This subfolder contains a randomly named JSON file. It contains just one record, which shows 175 MaxNewCasesReported in North America.

40. Switch back to the previous browser tab and wait 5 minutes.

    The chart on the Metrics page will display two more output events (for a total of three), and the output file will contain two more JSON objects that are exactly the same as the first one. In fact, because no more events were received — that is, you did not upload any more files — the output did not change, and events from the ABW file fell out the hopping window you specified in the query after three minutes.

41. Open the **/Daily** folder and the **globaldata** container. Then locate the **USA** file. (This file is also available in the companion content.)

    > **NOTE** Because Azure Storage Explorer uses pagination when displaying the folder content, you might need to navigate to the second or third page to find the file.

42. Copy (or upload) the file to the **InputData** folder in the container you have chosen as the BlobInput.

43. Wait for about two minutes. (Be sure you don't hit the three-minute mark.) Then open the **/Daily** folder and the **globaldata** container and locate the **ALB** file. (This file is also available in the companion content.)

44. Copy (or upload) the file to the **InputData** folder in the container you have chosen as the BlobInput.

45. Wait for five minutes. While you do, feel free to look at how the chart and the output file change.

    The monitoring chart displays two more spikes: 275 events (the USA file) and 188 events (the ALB file). Output events increment, too. In particular, one or two spikes (depending on how long you waited between the upload of the USA and ALB files) report two output events, whereas all other spikes report one output event.

    The output file now has more records in it. You will see three more records for North America, reporting 78427 MaxNewCasesReported, and three more records for Europe, reporting 178 MaxNewCasesReported. Of these six new records, two or four of them (depending on how long you waited between the upload of the USA and the ALB files) will report the same value in the OutTimes property. USA records will fall out of the hopping window before ALB ones. In fact, the last one or two records in the output files will report Europe in the Continent property.

46. Play around with other input files, uploading them into the InputData folder and watching how the chart on the Metrics page and the output change.

47. When you are finished, return to your resource's Overview page in the Azure Portal and click the **Stop** button.

Summary

- Modern data architectures contain a mixture of different workload types.

- A batch workload is a massive transformation of data usually from relational stores or data lakes. This data can fall under the domain of big data. Batch processing may require a specialized engine, like Hadoop-based systems or MPP platforms, which are able to scale as needed.

- ETL and ELT are two common patterns for data extraction and preparation. They share the same conceptual steps but end up with different implementations. In ETL, you use a specialized engine to perform the transform phase and then load data into a target repository. In ELT, the target repository also has the ability to transform data in a very effective way and move only the necessary data in it, so the load and transform phases are performed inside the target repository itself. ETL is more common in on-premises scenarios, and ELT is more common in hybrid or cloud scenarios.

- Azure Data Factory is a platform for performing data movement and data-processing orchestration. It has a visual authoring tool that you can use to create, manage, schedule, and monitor pipelines.

- A pipeline is a logical grouping of activities. Activities can be used to perform data movement at scale or to submit jobs to external services, like Azure HDInsight, Azure Databricks, Azure SQL Database, Azure Synapse Analytics, and so on.

- Datasets are objects used to represent data on a remote data store, and data stores are connected through linked services.

- Integration Runtimes (IRs) are the core of Azure Data Factory. They enable communication between linked services, datasets, and activities.

- Azure Data Factory has out-of-the-box connectors to the most popular services on the market, both on-premises and in the cloud. Access to sources not facing the internet can be gated through a particular type of IR called a self-hosted IR.

- Other IRs include the Azure IR and the Azure SSIS IR. The former is the default IR. The latter can be used to run SQL Server Integration Services packages in an on-demand cloud environment.

- Mapping data flows are Azure Data Factory components that can be used to author complex data transformations in a visual way. Behind the scenes, mapping data flows are translated into Spark code that is submitted to an on-demand Azure Databricks cluster.

- Pipelines can be invoked manually in the authoring portal, through a rich set of APIs (.NET, REST, and Python), or through triggers. Triggers can be bound to multiple pipelines and a single pipeline can be invoked by multiple triggers. The only exceptions to this are tumbling window triggers, which can be bound to only one pipeline at time.

- Triggers in Azure Data Factory can be of three types: schedule, tumbling window, and event-based.

- Azure Databricks is a Spark-based platform for massive data processing. It enables data engineers, data scientists, ML engineers, and data analysts to work together in a collaborative way.

- Spark is an open-source project that processes high volumes of data in memory. Databricks was founded by the original creators of Spark, and its platform is based on an optimized, closed distribution of Spark.

- Azure Databricks has native integration with the most popular Azure data services, such as Azure storage, Azure SQL Database, Azure Synapse Analytics, and Azure Cosmos DB.

- Spark supports many programming languages, including Scala, Python, R, SQL, and Java. Code can be entered through notebooks, which consist of independent cells that share the same context. You can access variables defined in previously executed cells.

- Thanks to Spark's versatility and maturity, Azure Databricks can be used in many scenarios, such as streaming, batch processing, ML, and data graphing. In addition, Delta Lake technology can be used to perform ACID operations on data lakes.

- To execute Spark code in a Databricks workspace, you must first create at least one cluster. A cluster is composed of a driver node and at least one executor. If autoscaling is enabled for the cluster, the engine will provision the minimum number of executors when the cluster starts and will spin up other executors only when a resource-intensive job is run. If the auto-shutdown option is enabled, additional executors are shut down after a specified period of inactivity.

- You can have multiple clusters in your workspace, each of them tailored to support a specific workload. In fact, you can have different combinations of runtime version, Spark version, number, size of executors, and so on.

- Premium-tier workspaces support role-based access control (RBAC) for notebooks, clusters, jobs, and tables. You can create groups to better manage security and user privileges, assigning to specific access rights to these groups (and not to individual users).

- Spark operates on data through objects called data frames. A data frame is based on Resilient Distributed Datasets (RDDs), which are immutable pointers to data stored on distributed or external data stores.

- Data frames can manipulate data through transformations, which are object methods that represent logical manipulations of the source data. Typical transformations include filters, aggregations, joins, and the selection or insertion of columns. As with external tables in Azure HDInsight, data is not accessed when a transformation is applied.

- Data frames physically access the data when actions are invoked. Spark computes an execution plan that considers all the applied transformations and submits a job to the cluster's driver node. Typical actions include displaying the data and writing it to a sink data store.

- Azure Synapse Analytics is an analytics service that offers data integration, enterprise data warehousing, and big data analytics, all in the same platform.

- When you provision an Azure Synapse Analytics resource, a workspace is created.

- An Azure Synapse Analytics workspace comes with a built-in serverless SQL pool that can be used to perform quick data exploration and non–mission critical workloads.

- You can create any number of dedicated SQL and Spark pools. The former are the well-known Azure SQL Data Warehouse (with a new name and integrated in the Azure Synapse Analytics workspace), whereas the latter are based on Microsoft's proprietary version of the open-source Apache Spark project.

- Traditional SQL Server tables can be created only in a dedicated SQL pool, whereas external tables can be created in both dedicated and serverless SQL pools.

- Azure Synapse Analytics can leverage PolyBase to access data that resides in external storage. PolyBase leverages the parallel processing capabilities of the Azure Synapse Analytics engine to scale out.

- After you set up PolyBase, you can use external tables in the same way you do in Hive. In this case, you use T-SQL to query and manipulate the data.

- PolyBase is one the fastest possible ways to load data into Azure Synapse Analytics. A well-known practice is to use a staging table (usually a heap) as an intermediate step between the source data and the destination table.

- The OPENROWSET and COPY commands allow for a fresh and flexible yet scalable approach to load CSV and Parquet files in and out of Azure Synapse Analytics, introducing, for example, schema discovery when parsing the file content.

- Azure Databricks and Azure Data Factory can leverage PolyBase when the sink of a data movement is Azure Synapse Analytics.

- Working with Spark in Azure Synapse Analytics is very similar to working with it in Azure Databricks, although integration between Spark and dedicated SQL pools is even stronger.

- A stream workload involves the real-time processing of messages (or events) produced by sensors, devices, and applications. These messages are usually aggregated by time windows and sent to live dashboards. Raw messages are stored in high-capacity and low-cost stores like data lakes for further analysis.

- Azure Stream Analytics is the platform as a service (PaaS) stream-processing engine in Azure.

- Azure Stream Analytics is tightly integrated with stream-ingestion engines like Azure Event Hubs (with or without the Kafka interface) and Azure IoT Hubs. In addition, it can monitor storage accounts for events like blobs being created, updated, or deleted.

- An Azure Stream Analytics job is composed of one or more inputs, one or more outputs, and a query.

- Inputs are divided into two categories: data streams and reference data. The former are streams of events ingested through Azure Event Hubs, Azure IoT Hubs, or Blob storage. The latter are static (or almost static) data stored on an Azure SQL Database or a Blob storage. Reference data inputs are used to enrich the events in the data streams. At least one data stream must be present.

- Many services are supported as target outputs. Azure Event Hubs, Azure IoT Hubs, Blob storage, Azure SQL Database, Azure Cosmos DB, and Power BI are probably the most common ones. In particular, Power BI can be leveraged to build real-time dashboards over the processed streams.

- Queries process input events in real time (or near real time) to produce output results. The query language has a SQL-like syntax but provides additional built-in functions to process batches of events all at once.

- Built-in functions are divided into the following categories: aggregate, analytic, array, geospatial, input metadata, record, windowing, and scalar.

- Before being aggregated or transformed, events are usually grouped by time windows based on their arrival time. The most common types of windowing functions are hopping, sliding, tumbling, and session.

- A query can read from multiple inputs at once, joining them to create complex data streams or to enrich events in transit with reference data.

- A query can write to more than one output at once. To do so, you need to write multiple statements, specifying the target output with the INTO clause.

- An Azure Stream Analytics job consumes Streaming Units (SUs). You can scale the number of SUs available for the job up or down, from a minimum of 1 to a maximum of 192. Autoscaling is not supported.

Summary exercise

In this summary exercise, test your knowledge about the topics covered in this chapter. You can find answers to the questions in this exercise in the next section.

Adventure Works, a bicycle company, wants to build customer loyalty. To do so, it has developed a mobile app that tracks customer wellness data, ingesting it in real time to Azure Event Hubs through REST calls over HTTPS. At the moment, data is simply offloaded automatically to a Blob storage account in Avro format using the Event Hubs Capture feature and no further analysis is done on it.

The IT department has been asked to develop three processes to extract value from all the data the company is gathering:

- A process that feeds a real-time dashboard in Power BI
- A daily process that consolidates all the collected information in a data warehouse
- A daily process that makes all the raw events offloaded by the Event Hubs Capture feature available to data scientists

There are some constraints to consider when architecting these processes:

- All processes must be cloud-only and hosted on Microsoft Azure.
- The data scientists are used to Spark, and they are eager to leverage the collaborative development experience offered by Databricks. So, Azure Databricks must be in the picture.

- The data scientists are fine dealing with the Avro format used by Event Hubs Capture. They will take care of all the parsing needed.

- Platform as a service (PaaS) offerings are preferred over infrastructure as a service (IaaS) whenever possible. When a cluster is needed for any reason, it's preferable to have the option to use an on-demand cluster rather than opting for an always-on cluster.

- The data warehouse will be hosted on Azure Synapse Analytics. The daily process that consolidates the data must be optimized to target this service.

- The data-engineering team does not like to use a graphical approach to developing ETL/ELT pipelines. This team requires full control over the code that performs the needed transformations over the data. The only exception to this is when no transformations are involved — for example, when data can be moved as is from source to destination.

As the company's IT manager, it's your job to evaluate Azure services to pick the best ones for your needs. Considering the scenario described, answer the following questions:

1. How will you develop the real-time process?

   A. Use Azure Data Factory to set up a pipeline that will move Event Hub Capture Avro files to a dataset in Power BI (transforming the data accordingly) and schedule it to run every minute.

   B. Use Azure Stream Analytics to set up a job with Azure Blob storage as the input and Power BI as the output. As soon as a new file is placed in the Event Hub Capture folder, a query will output the aggregated and transformed events to a Power BI dataset.

   C. Use Azure Stream Analytics to set up a job with Event Hubs as input and Power BI as output. As soon as events are ingested into Event Hubs, a query will output the aggregated and transformed events to a Power BI dataset.

   D. Use Azure Databricks and Spark streaming to aggregate and transform events ingested into Event Hubs, outputting the results to a Power BI dataset. Although this solution will require an always-on cluster, it is the only practical option.

2. How will you develop the daily process that will consolidate data in Azure Synapse Analytics?

   A. Data transformation and loading will be performed in Spark inside notebooks in Azure Databricks. PolyBase will be leveraged through the Azure Synapse Analytics connector offered by Azure Databricks. A dedicated Spark cluster will be created with auto-shutdown option enabled. The process will start the cluster on-demand; when the process ends and after the specified timeout, the cluster will shut itself down automatically. Azure Data Factory will orchestrate the whole process.

   B. Data transformation and loading will be performed in Spark inside notebooks in Azure Databricks. Because the Azure Synapse Analytics connector offered by Azure Databricks does not support PolyBase, data loading in Azure Synapse Analytics will be performed in batches, although it will be slower. The process will start the cluster on-demand; when the process ends and after the specified timeout, the cluster will shut itself down automatically. Azure Data Factory will orchestrate the whole process.

C. Data transformation and loading will be performed in Azure Data Factory with mapping data flows using an elegant code-free approach. PolyBase will be leveraged through the Azure Synapse Analytics connector offered by Azure Data Factory. The process will run on an on-demand, pay-as-you-go Azure Databricks cluster, fully managed by Azure. Azure Data Factory will also orchestrate the whole process.

3. How will you develop the daily process that will make Event Hubs Capture data available to the data scientists?

A. Data movement will take place in Azure Data Factory using a simple pipeline with a copy activity inside it.

B. Data movement will be performed by a pass-through query in Azure Stream Analytics.

C. Data movement will be performed by a Spark notebook in Azure Databricks.

Summary exercise answers

This section contains the answers to the summary exercise questions in the previous section and explains why each answer choice is correct.

1. **C: Use Azure Stream Analytics to set up a job with Event Hubs as input and Power BI as output**   This is a typical real-time scenario, and Azure Stream Analytics meets all the requirements. It integrates very well with Azure Event Hubs and Power BI and can scale up to accommodate resource-intensive streams. Although option B also uses Azure Stream Analytics, it would not be a true real-time process because it depends on the offloading of events performed by Event Hubs Capture technology. (This is true also for option A.) Option D could be a viable solution, but taking into consideration the project constraints, option C is preferable.

2. **A: Data transformation and loading will be performed in Spark inside notebooks in Azure Databricks**   Azure Databricks fits perfectly in this scenario and will grant optimal performance in the load phase thanks to PolyBase support. Option C would work too, but the code-free approach would not be acceptable to the data-engineering team. In fact, data transformations would be likely needed to prepare data for Azure Synapse Analytics.

3. **A: Data movement will take place in Azure Data Factory using a simple pipeline with a copy activity inside it**   A copy activity in Azure Data Factory is one of the cleanest ways to perform data movement at scale in Azure. Azure Stream Analytics is a stream-processing engine and is not meant for batch scenarios, so option B must be discarded. Option C could be viable, but because this is a simple data movement from source to destination (that is, transformations and file-format conversions are not required) a code-free approach is the best solution and would not be affected by any constraint.

## Chapter 4

Monitor and optimize data solutions

Any resource deployed in Azure will work under normal conditions as expected. However, bad things can happen over time. Even if there is no problem at all, your best estimations will not exactly match reality in terms of resource consumption, data volume, and usage.

You will need to monitor what is going on with your resources — how they are used, how well they react under stress — and take actions to address malfunctions or optimize behavior and resource usage.

All Azure resources start to send information to Azure Monitor as soon as they are deployed. Azure Monitor is a centralized service to collect resource data. It receives two kinds of information:

- **Metrics**   These are small pieces of data, almost always numeric, with information about resources like memory usage, CPU consumption, space available or in use, amount of data transferred, and so on. This information is stored in time series format. It is almost instantaneous and can be used to define alerts, KPIs, and the like.

- **Logs**   Each service tracks information about its activities, which is stored in a repository, or log. These logs have different formats depending on the service and the level, but all of them can be queried to search for and review specific entries.

Azure uses these to show you information about each resource in the resource's Overview page. You can refine this information or add more detailed reports as desired. You can also create new dashboards in Azure Portal to make your most important information readily available. And you can query this information and prepare your own reports with Power BI, Jupyter Notebooks, or other reporting tools.

> **NOTE**   With Jupyter Notebooks, information from Azure Monitor can be reported in a graphical style with any notes or explanations you want to add.

This chapter analyzes different tools for obtaining information from Azure Monitor, the principles of the monitoring process in Azure Monitor, and how to use this information to optimize your solutions.

Topics covered in this chapter:
- Monitor data storage
- Monitor data processing
- Optimize Azure data solutions

Monitor data storage

Monitoring data storage is about keeping control of the storage space used, the number of reads and writes, and the relationship between usage and storage service level to evaluate whether resources are underused or overused.

> **NOTE**   Different kinds of storage require different monitoring approaches.

Monitor an Azure SQL Database

Azure SQL Database is part of the Microsoft SQL Server family. As such, it has very fine-tuned internal monitoring capabilities. Moreover, these monitoring processes are evolving, and will likely become even better over time.

> **NOTE**   Later in this section, you will learn how to add custom monitoring procedures, which you can use instead of the ones that are integrated in the product.

> **NOTE**   Data from these monitoring processes is stored in the Azure Monitor metrics.

The Overview page of any Azure SQL Database in the Azure Portal displays three monitoring graphs by default: Compute Utilization, App CPU Billed, and Database Data Storage.

Figure 4-1 shows the first of these, Compute Utilization. This graph presents the database's compute usage. It shows three different maximum values for different slices of time:
- CPU percentage
- Data IO percentage
- Log IO percentage

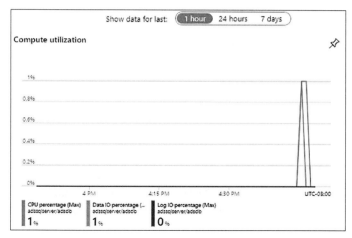

FIGURE 4-1 Azure SQL Database Compute Utilization graph

The three boxes along the bottom of the graph contain the values for these three metrics. If you move the mouse pointer over the graphic, these change to reflect more exact values. A filter tool along the top of the graphic allows you to select which period you want to display, between 1 hour, 24 hours, and 7 days.

Figure 4-2 shows the App CPU Billed graph. An Aggregation Type drop-down list at the top of the graph enables you to select the type of value: Max, Min, or Avg. Notice that this setting applies not only to this graph, but to the Compute Utilization graph, too. The filters along the top of the Overview page are for the complete set of information.

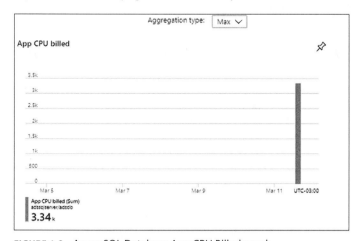

FIGURE 4-2 Azure SQL Database App CPU Billed graph

Figure 4-3 shows the third graph on the Overview page: Database Data Storage. This is a donut chart that shows the total space in the database as well as the space currently in use.

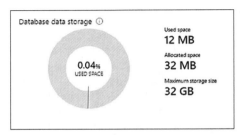

FIGURE 4-3 Azure SQL Database Database Data Storage graph

> **NOTE** Figures 4-1, 4-2, and 4-3 were taken just after we "woke" the database by executing a couple of queries, since it is configured to pause after one hour of no utilization.

Metrics

For Azure SQL Database, metrics are taken directly from the SQL service to enable you to measure how the service and the database behave.

There could be different namespaces for metrics, grouped to facilitate selection. But in the case of an Azure SQL Database, there is only one namespace: SQL Database standard metrics. Inside the namespace, however, metrics can be grouped. Azure SQL Databases have a Basic group and an Instance and App Advanced group.

> **NOTE** Other resources use different namespaces and groups.

Table 4-1 lists the most important metrics in the SQL Database standard metrics namespace.

TABLE 4-1 Important metrics in the SQL Database standard metrics namespace

Metric Group	Metrics
Basic	App CPU Percentage
	App Memory Percentage
	Blocked by Firewall
	CPU Limit
	CPU Percentage
	CPU Used
	Data IO Percentage
	Data Space Used Percent
	Deadlocks
	Failed Connections
	In Memory OLTP Storage Percent
Instance and App Advanced	SQL Server Process Core Percent
	SQL Server Process Memory Percent

PRACTICE   Create a custom chart and dashboard for metrics

You can create your own graphs and place them on your standard dashboard or on a custom one. This gives you a way to follow the metrics that matter most to you. In this practice exercise, you will create a graph of certain metrics and add it to a dashboard.

1. Open *https://portal.azure.com* and use your credentials to log in.

2. Navigate to **Azure SQL Database**.

3. On the **Overview** page, scroll down in the left pane to the **Monitoring** section and click the **Metrics** option.

    A metrics chart editor page opens. You use this to create new metrics. (See Figure 4-4.)

FIGURE 4-4   Metrics chart editor

4. Click the pencil to the right of the **Title** box and, for this example, type **Security**.

5. Click the **Metric** drop-down list and choose **Blocked by Firewall**.

6. Click the **Aggregation** drop-down list and choose **Count**.

7. Click the white **check mark** with a blue background that appears to the right of the **Aggregation** setting or simply click in the graph area to create the graph.

8. In the toolbar along the top of the page, click the **Chart Type** drop-down list (which in Figure 4-4 has Line Chart selected) and choose **Area Chart**.

    Now let's add the new chart to your database dashboard.

9. Click the **Pin to Dashboard** button in the toolbar along the top of the page.

10. In the panel that appears, click the **Create New** tab and, for this practice exercise, type **ADS My Dashboard**.

11. Leave the **Private** setting as is.

12. Click the **Create and Pin** button at the bottom of the panel.

13. Click the **Dashboard** option at the top of the page to return to the database Overview page.

14. Click the **My Dashboard** drop-down list and choose **ADS My Dashboard**.

15. To auto-refresh the chart, click the **Auto Refresh** button and choose a frequency value from the list that appears. (See Figure 4-5.)

FIGURE 4-5 Dashboard auto-refresh options

Logs

Logs are another important information source. For example, there are logs to store information about the SQL Server's or SQL Database's security, service health, resources health, policies, and so on.

> **NOTE** These logs are for the service and/or database, not for the data, and are not related to the database transaction log.

One useful log is the Recommendation log, which collects entries from the Azure Maintenance service. The Recommendation log measures resource conditions and automatically provides recommendations to optimize them.

Logs can be archived in a storage account, sent to Azure Log Analytics, or sent as a stream to an Azure Event Hub.

Monitor Azure SQL Database using DMV

As with on-premises SQL Server deployments, you can evaluate what is happening within your database from inside it. This is useful when you need to identify which database operations are consuming resources or blocking processes. You use Dynamic Management Views (DMVs) to view this information.

Unless the connected user is a server admin or database owner, special permissions must be granted to enable the querying of DMVs. To grant this, issue the following T-SQL statement:

```
GRANT VIEW DATABASE STATE TO <user>
```

Table 4-2 lists the DMVs you can query (with the proper rights) in an on-premises SQL Database or in Azure SQL Database, categorized by type.

TABLE 4-2 DMVs

DMV Type	DMVs

Views	`sys.dm_db_file_space_usage`
	`sys.dm_db_fts_index_physical_stats`
	`sys.dm_db_log_space_usage`
	`sys.dm_db_partition_stats`
	`sys.dm_db_persisted_sku_features`
	`sys.dm_db_session_space_usage`
	`sys.dm_db_task_space_usage`
	`sys.dm_db_uncontained_entities`

Functions	`sys.dm_db_log_info`
	`sys.dm_db_log_stats`
	`sys.dm_db_page_info`

Azure SQL Database–specific	`sys.dm_database_copies (only in master database)`
	`sys.dm_db_objects_impacted_on_version_change`
	`sys.dm_resource_stats`
	`sys.dm_db_resource_stats`
	`sys.dm_db_wait_stats`
	`sys.dm_operation_status`

Geo-replication	`sys.dm_continuous_copy_status`
	`sys.dm_geo_replication_link_status`
	`sys.dm_operation_status`
	`sys.geo_replication_links`

NEED MORE REVIEW? For more information about querying DMVs and to see some samples, visit *https://docs.microsoft.com/en-us/azure/azure-sql/database/monitoring-with-dmvs*.

Intelligent Insights

Considering how much information might be available, it can be difficult to find the information you need about an issue in your database. To help with this, Azure teams implemented Intelligent Insights. This feature compares different values from DMVs with the same metrics during the last seven days, looks for significant differences, and flags them.

Intelligent Insights is implemented at the workspace level and must be added from the Solution Gallery. Once deployed, it collects information from all your databases. It uses all metrics and logs you configure and send using diagnostic settings. For more information about Intelligence Insights, including sample cases, see *https://docs.microsoft.com/en-us/azure/azure-sql/database/intelligent-insights-overview*.

> **NOTE** Intelligent Insights is in preview at the time of this writing.

Monitor Blob storage

Azure storage accounts, including Azure Blob storage accounts, record several metrics as soon as they are created. Because each account manages different resources, there are different namespaces for them, including one for the entire account. When you view an account's Overview page, it shows four generic charts for the entire account by default: Total Ingress, Total Egress, Average Latency, and Request Breakdown.

Metrics

If you want to track metrics for your Azure Blob storage, you can create your own charts. The steps are similar to the ones you followed in the "Create a custom chart and dashboard metrics" practice exercise in the "Monitor an Azure SQL Database" section. You just change the storage account's namespace to blobs.

As shown in Table 4-3, there are two types of metrics for Azure Blob storage: capacity and transaction.

TABLE 4-3 Metrics for Azure Blob storage

Metric Type	Metrics	Description
Capacity	Blob Capacity	The total used space measured in bytes.
	Blob Container Count	The number of containers defined for the account.
	Blob Count	The total number of items stored in the account in all containers.
	Index Capacity	The amount of space used by indexes in containers. This counter is important for data lake storage and is discussed in more detail in the next section.
Transaction	Availability	The availability of the service, in percentage form.
	Egress	The number of bytes sent out to external connections and to other Azure services.
	Ingress	The number of bytes received from external connections and other Azure services.
	Success E2E Latency	The time required to send required content or store received data. This includes all the steps, like searching, gathering data, and replying to the request. It is usually aggregated as an average.
	Success Server Latency	Like Success E2E Latency, but excluding network latency. This metric evaluates only the latency generated by the storage.
	Transactions	All requests to the blob, sent and received, with no distinction between successful requests, errors, and fails.

Filtering metrics data

Most of these metrics are a consequence of different actions or APIs connecting to the Azure Blob storage. Sometimes, you will need to refine your observations about what is going on. For that, you have the filter feature.

In the toolbar along the top of the Metrics page is an **Add Filter** button. Click this button to display the Filter Editor. Here, you can select a property type, the operator to apply to the filter, and the specific value to filter for. Table 4-4 lists the filter options for blobs.

TABLE 4-4 Filter options for blobs

Filter Type	Description	Examples
API Name	Defines the kind of action executed against the blob.	Delete File Get Blob List Blobs List Containers Put Blob
Authentication	Identifies the kind of security used by the actions performed.	Account Key Anonymous OAuth SAS
Geo Type	Used to select a location to evaluate if geo-replication is configured for the storage. If not, Primary is the only available option.	[Location] Primary

Another type of filter is the Period filter. It is the right-most button on the toolbar along the top of the page. You can click it to view options for selecting time ranges — for example, the last 5 minutes or the last 300 days. There is also a Custom option, which enables you to precisely define a start and end date and time, as well as the time granularity. By default, the estimated time granularity is automatically calculated depending on the data. However, you can change it to a value ranging from 1 minute to 1 month. (See Figure 4-6.)

FIGURE 4-6 Period filter options

In addition to filtering metrics data, you can choose to display this data in different forms, including line graphs, bar graphs, area graphs, scatter points, and grid view. Note that grid view simply displays a grid of values, not a graph, for the available aggregations for the metrics. See Table 4-5 for an example.

TABLE 4-5 All blob counters in a grid view

Metric	Avg	Sum
Blob Capacity	17.4 MiB	N/A
Index Capacity	436.8 KiB	N/A
Blob Count	212	N/A
Availability	100 %	N/A
Success E2E Latency	94.61 ms	N/A
Success Server Latency	22.17 ms	N/A
Blob Container Count	2	N/A
Egress	N/A	299.9 KiB
Ingress	N/A	18.9 KiB
Transactions	N/A	29

Logs

Any activity performed in a storage account is logged automatically. To see a list of the most recent activities, click the **Activity Log** option in the left panel of any storage account page in Azure Portal.

These activities are just for the account itself, not for its contents. For example, you will often see entries for a List Storage Keys activity, which is logged each time a user connects to the storage account or reviews the account's storage keys. All activities are registered twice: once when the activity starts, and once when it completes or fails. (See Figure 4-7.)

Operation name	Status	Time
∨ ⓘ List Storage Account Keys	Succeeded	2 hours ago
ⓘ List Storage Account Keys	Started	2 hours ago
∨ ⓘ List Storage Account Keys	Succeeded	6 hours ago
ⓘ List Storage Account Keys	Started	6 hours ago

FIGURE 4-7   Activity log entries

You can define other logs to be updated by any of your resources. These become part of the information managed by Azure Log Analytics. Later in this chapter, you will learn how to use Azure Log Analytics to query and manage log data.

Implement Azure Data Lake Storage monitoring

Because data in Azure Data Lake Storage (ADLS) is stored in blob containers, monitoring ADLS essentially involves monitoring a blob container in a storage account along with the Azure storage account for general metrics. However, there is a special metric within the blob metrics that is specifically for ADLS.

Recall that ADLS requires you to enable the hierarchical namespace, which maintains indexes of the hierarchy, to accelerate data retrieval. So, the Index Capacity metric mentioned earlier informs you how much space is used by indexes in the containers. This metric, when compared with the Blob Capacity metric, could flag a bad hierarchy design if the index is a large percentage of the entire blob capacity. In other words, the index capacity should be only a small percentage of the entire blob capacity because your hierarchy should not incur a high data cost.

Implement Azure Synapse Analytics monitoring

Azure Synapse Analytics, like other storage resources, tracks several metrics. However, as with Azure SQL Database, these metrics are grouped under the SQL Database standard namespace.

This does not mean they are the same, however. Some metrics in Azure SQL Database *are* present in Azure Synapse Analytics, but others are unique to each resource. For example, Azure Synapse Analytics has several DWU counters, like DWU Limit, DWU Percentage, and DWU Used, which make no sense in Azure SQL Database.

Azure Synapse Analytics also has a set of Workload Management metrics, which make it easy to see what happens with resource usage during the data processing. These metrics enable you to assign resources, levels of importance, and specific resource-management metrics to certain workloads.

The most important Workload Management metrics are as follows:

- **Effective Cap Resource Percent**   Reveals the percentage of resources currently in use by a workload. Having several workloads running in parallel requires the distribution of resources, which will be based on the importance level of each workload. You can establish a minimum percentage assigned to a workload (see the next bullet). This reduces the effectiveness of others.

- **Effective Min Resource Percent**   The minimum percentage of resources assigned to a workload. This ensures that at least this percentage of resources will be reserved for the workload. If the Effective Min Resource Percent metric is very low compared to the Effective Cap Resource Percent metric described in the preceding bullet, you will probably want to adjust it to enhance workload resource utilization and behavior.

- **Workload Group Active Queries**   The number of queries running in a workload. More queries generally require more capacity. In some cases, you can use this metric to evaluate whether some queries in a workload are consuming excessive resources. If so, it might be that they need to be moved to other workloads, or that the current workload just needs a higher Effective Min Resource Percent setting.

- **Workload Group Allocation by Max Resource Percent**   The proportional resource utilization in percentage form. Reaching a value of 100% means that under pressure, the workload will need more resources than assigned.

- **Workload Group Allocation by System Percent**   The percentage of the resources for an entire service used by a particular workload.

Comparing the Workload Group Allocation by System Percent, Effective Cap Resource Percent, and Effective Min Resource Percent metrics can provide useful insights. For example, it can help you identify the best Effective Min Resource Percent value — one that is closer to the average allocated by the system — and help you optimally distribute resources among workloads. In addition, combining the Workload Management metrics with the standard metrics for the entire service — like CPU or memory percentage — enables you to fine-tune resource utilization.

NEED MORE REVIEW?   For more information about workload management, see *https://docs.microsoft.com/en-us/azure/synapse-analytics/sql-data-warehouse/sql-data-warehouse-workload-management*.

Implement Cosmos DB monitoring

Cosmos DB includes metrics for most activities on the platform. However, because Cosmos DB has different APIs and is based on a geo-replication implementation, metrics evaluation is slightly different for this database.

To facilitate a better understanding of the metrics for Cosmos DB, the database's Metrics page in the Azure Portal shows a custom dashboard that displays the most important metrics in various tabs. (See Figure 4-8.)

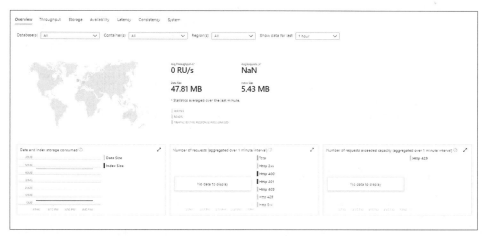

FIGURE 4-8 A custom dashboard containing the most important metrics for a Cosmos DB account

Table 4-6 describes the various charts available from each tab in the database's Azure Portal page.

TABLE 4-6 Cosmos DB charts

Tab	Charts	Description
Overview	Global Stats and Geo-Distribution	Displays a map with dots representing the regions where the Cosmos DB account is replicated. This chart also displays RU and request averages, the size of the data stored, and the indexes.
	Data and Index Storage Consumed	A line chart that shows the size evolution of data and indexes.
	Number of Requests	A line chart with different lines for the HTTP codes returned by the account, including the HTTP-200 code family, the HTTP 400 code family, and so on.
	Number of Requests Exceeded Capacity	A line chart that presents the number of exceeded requests, with different lines for different HTTP response codes.
Throughput	Number of Requests	A line chart with different lines for the HTTP codes returned by the account, including the HTTP-200 code family, the HTTP 400 code family, and so on.
	Number of Requests Exceeded Capacity	A line chart that presents the number of exceeded requests, with different lines for different HTTP response codes.

Tab	Charts	Description
Storage	Data and Index Storage Consumed	A line chart that shows the size evolution of data and indexes.
Availability	Availability	A line chart that represents the percentage of the availability of the service versus the SLA.
Latency	Read Latency	A line chart that shows end-to-end read latency compared to the SLA.
	Write Latency	A line chart that shows end-to-end write latency compared to the SLA.
Consistency	Consistency	A chart that shows how the configured consistency is working, based on three values: probability of strongly consistent reads based on your workload, replication latency trend, and time after commit (ms) in read region.
System	Number of Metadata Requests	A line chart with the number of metadata requests, with different lines for different HTTP response codes.
	Number of Metadata Requests Exceeded Capacity	A line chart with the number of metadata requests that exceeded capacity, with different lines for different HTTP response codes.

Configure Azure Monitor alerts

Of course, having metrics available is of little use if you must be on the site to see what is going on every minute. That's where alerts come in. In Azure, you can define alerts on any metrics value. These link to the platform's alert infrastructure, which is supported by Microsoft Insights Provider.

Alerts prevent failures by communicating when a metric's value falls outside an acceptable range. For example, if you set an alert for the CPU percentage for an Azure SQL Database, and some event triggers that alert, you can investigate the reason of the alert and solve the problem or assign more resources before a failure.

PRACTICE   Enable Azure Insights Provider

You must associate Azure Insights Provider with the subscription to enable alerts for it. This practice exercise shows you how.

1. Open *https://portal.azure.com* and use your credentials to log in.
2. In the search box, type **subscriptions**, and select the desired subscription from the drop-down list that appears.
3. In the **Settings** group in the left panel, click **Resource Providers**.
4. Type **insights** in the **Filter by Name** box.
5. Select **microsoft.insights** in the list that appears and click the **Register** button along the top of the page. (Notice it could be already marked as Registered, so there is nothing else you need to do.)

PRACTICE   Define an action group

Microsoft Insights Provider uses action groups to send information and alerts. So, you will need to define action groups before creating alerts on metrics. All alerts must be assigned to one or more action groups. Follow these steps to define an action group to receive alerts:

1. Open *https://portal.azure.com* and use your credentials to log in.

2. In the search box, type **alerts**, and select the corresponding entry from the drop-down list that appears.

3. Click the **Manage Actions** button along the top of the **Alerts** page.

4. In the **Manage Actions** page, click **Add Action Group**.

5. In the **Basics** tab of the **Create Action Group** page, configure the following settings:

   - **Subscription**   Confirm that the selected subscription is the correct one.

   - **Resource Group**   Choose the resource group to which the action group will belong.

   - **Action Group Name**   Type a name for the new action group. This will be the internal name. For this practice exercise, type **ADSActionGroup**.

   - **Display Name**   Enter a display name for the action group. This should be a more user-friendly name than the action group name. In this case, type **ADSAlerts**.

6. Click the **Review + Create** tab. Then click the **Create** button.

> **NOTE**   You can define action groups to take actions instead of simply sending alerts. This enables you to automatically manage certain situations, like resource overloads, failures, and so on, making the entire implementation more responsive and fault-tolerant.

Create alerts for metrics

When you need to be notified about potential problems, you can define alerts for any metric in the metrics chart editor. Based on your alert definition, Azure will process any event that triggers the alert and will take the action assigned.

> **NOTE**   You can define alerts and other automatic actions when you create a new action group, or you can add them to an existing action group in the Manage Actions page. It is important to plan your action groups to effectively reduce risk and failures.

EXERCISE 4-1   Create an alert from the Azure Portal

In this exercise, you will learn how to create an alert and see what happens when it is triggered. You will do this by creating a fake alert, setting a very low threshold for CPU percentage and forcing it to trigger with some simple T-SQL code.

1. Open *https://portal.azure.com* and use your credentials to log in.

2. Navigate to your **Azure SQL Database**.

3. In the left pane, in the **Monitoring** group, click **Metrics**.

4. In the metric selector on the **metrics chart editor** page, open the **Metric** drop-down list and choose **CPU Percentage**.

5. Open the **Aggregation** drop-down list and choose **Max**.

6. In the toolbar above the metric selector, click the **New Alert Rule** button.

7. In the **Create Alert Rule** page, click the **Whenever the Maximum CPU Percentage Is Greater Than <logic undefined> %** link.

8. In the **Configure Signal Logic** panel, under **Alert Logic Title**, configure the following settings:

   ■ **Operator**   Confirm that this is set to **Greater Than**.

   ■ **Aggregation Type**   Set this to **Maximum**.

   ■ **Threshold Value**   Set this to **4**. (This is a very low value, for testing purposes.)

   Leave all other settings at their default.

9. Click **Done**.

10. In the **Actions** tab, click **Select Action Group**.

11. Under **Select an Action Group to Attach to This Alert Rule**, choose the action group you created earlier and click the **Select** button.

12. Type a name for the alert in the **Alert Rule Name** box — for example, **CPU Alert**.

13. Leave the remaining settings with their default values.

14. Click the **Create Alert Rule** button.

    When the alert has been created, you'll receive a confirmation notice. It can take as long as 10 minutes for the rule to be active.

15. In the left panel, click **Query Editor**, and connect to it.

    If the dataset is paused, it will be a couple of minutes before you are allowed to connect.

16. Execute the following script to create a test schema and table:

```
CREATE SCHEMA ADSCh4
AUTHORIZATION dbo
GO
select * into ADSCh4.SalesOrderHeader from SalesLT.SalesOrderHeader
```

17. After the test objects are created, run the script in Listing 4-1 to elevate CPU usage in the Sales Order Header table.

LISTING 4-1 Insert data into Sales Order Header table

```
DECLARE @adder INT = 1;

WHILE @adder < 5
  BEGIN
    INSERT INTO [ADSCh4].[SalesOrderHeader]
          (
          [SalesOrderID],
          [RevisionNumber],
          [OrderDate],
          [DueDate],
          [ShipDate],
          [Status],
          [OnlineOrderFlag],
          [SalesOrderNumber],
          [PurchaseOrderNumber],
          [AccountNumber],
          [CustomerID],
          [ShipToAddressID],
          [BillToAddressID],
          [ShipMethod],
          [CreditCardApprovalCode],
          [SubTotal],
          [TaxAmt],
          [Freight],
          [TotalDue],
          [Comment],
          [rowguid],
          [ModifiedDate]
          )
    SELECT
          [SalesOrderID] + @adder,
          [RevisionNumber],
          DATEADD(month, @adder, [OrderDate]),
          DATEADD(month, @adder, [DueDate]),
          DATEADD(month, @adder, [ShipDate]),
          [Status],
          [OnlineOrderFlag],
          [SalesOrderNumber],
          [PurchaseOrderNumber],
          [AccountNumber],
          [CustomerID],
          [ShipToAddressID],
          [BillToAddressID],
```

```
            [ShipMethod],
            [CreditCardApprovalCode],
            [SubTotal],
            [TaxAmt],
            [Freight],
            [TotalDue],
            [Comment],
            [rowguid],
            [ModifiedDate]
      FROM
            [ADSCh4].[SalesOrderHeader];
   END;
GO
```

After about 5 minutes, you will receive an alert via email.

18. To clean up, type **alerts** in the search box at the top of Azure Portal and select the corresponding entry from the drop-down list that appears.

    You will see an alert chart, with the recently activated alert.

19. Click the **Manage Alert Rules** button in the toolbar along the top of the page.

20. Select the check box next to the alert you created and click the **Delete** button.

> **NOTE** As you saw in the last steps of this exercise, alerts are centralized for the entire subscription and can be established for several metrics and events from different resources.

Audit with Azure Log Analytics

Where metrics store data in time series, which are strictly numeric, logs could have several different data types and structures. Metrics enable you to see counters from your resources and how they vary during usage. With logs, you can analyze *why* metrics vary and diagnose possible issues in your implementation.

PRACTICE   Prepare an Azure Log Analytics workspace

To have a place to send all your logs, you must configure an Azure Log Analytics workspace. To do so, follow these steps:

1. Open *https://portal.azure.com* and use your credentials to log in.

2. In the search box, type **Log Analytics**, and select the corresponding entry from the drop-down list that appears.

   The Log Analytics Workspaces page opens.

3. Click the **Add** button.

4. In the **Basics** tab of the Create Log Analytics Workspace page that appears, configure the following settings:

   - **Subscription** Select your subscription.
   - **Resource Group** Select the resource group to which this workspace will belong.
   - **Name** For this practice exercise, type **ADSLogAn**.
   - **Region** Choose a region. (The resource group region is selected by default.)

5. Click the **Pricing Tier** tab and choose the desired tier. (The default is **Pay As You Go**.)

6. Click the **Tags** tab and enter any custom tags you like.

7. Click the **Review + Create** tab. Then click the **Create** button.

Enable server-level auditing

Your next step is to enable server-level auditing to track database events and write them to an audit log. You do this from the Azure SQL Server **Overview** page by clicking the **Auditing** tile. This will apply the auditing feature for its contained databases as well.

As shown in Figure 4-9, when you enable server-level auditing, you must select and configure at least one destination to store the audit log. Options include Azure Log Analytics, an Azure storage account, and Azure Event Hub.

> **NOTE** Logs can be stored in ADLS Gen2 storage but the logs do not support hierarchical namespaces, so the logs could not be shared with a Data Lake.

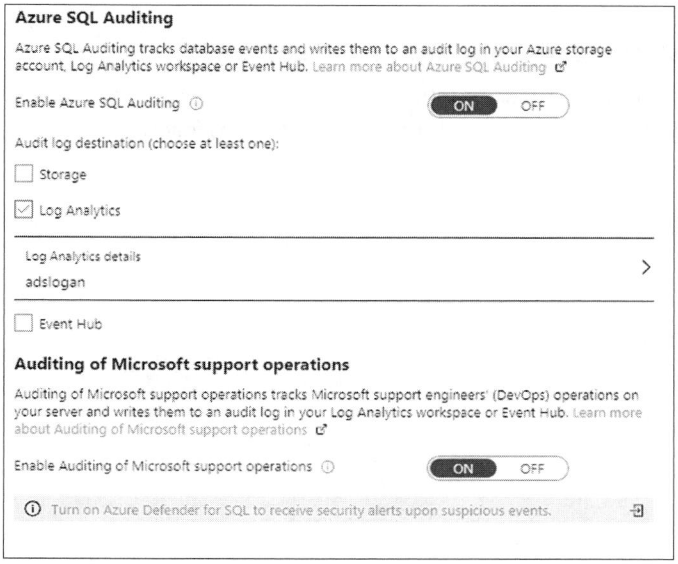

Azure SQL Auditing

Azure SQL Auditing tracks database events and writes them to an audit log in your Azure storage account, Log Analytics workspace or Event Hub. Learn more about Azure SQL Auditing

Enable Azure SQL Auditing ⓘ   (ON)   OFF

Audit log destination (choose at least one):

☐ Storage

☑ Log Analytics

Log Analytics details
adslogan   >

☐ Event Hub

Auditing of Microsoft support operations

Auditing of Microsoft support operations tracks Microsoft support engineers' (DevOps) operations on your server and writes them to an audit log in your Log Analytics workspace or Event Hub. Learn more about Auditing of Microsoft support operations

Enable Auditing of Microsoft support operations ⓘ   (ON)   OFF

ⓘ Turn on Azure Defender for SQL to receive security alerts upon suspicious events.

FIGURE 4-9 Azure SQL Auditing page

Configure logs

Of course, logs must be created before they can be analyzed. Some resources, like virtual machines (VMs), have predefined logs. Others, like applications, can be customized from within the application. For data sources, however, logs must be configured. Table 4-7 lists logs that can be configured for Azure SQL Database, organized into two groups: logs and metrics.

TABLE 4-7   Logs for Azure SQL Database

Log Type	Log
Logs	SQLInsights
	AutomaticTuning
	QueryStoreRuntimeStatistics
	QueryStoreWaitStatistics
	Errors
	DatabaseWaitStatistics
	Timeouts
	Blocks
	Deadlocks
Metrics	Basic
	InstanceAndAppAdvanced
	WorkloadManagement

NOTE   The metrics group allows you to store metrics in a log and to keep them for 90 days, which is the limit by default.

PRACTICE   Add diagnostic settings to a Cosmos DB account

For Cosmos DB, logs are generated based on diagnostic settings. These diagnostic settings are grouped by database API. They also include some RU consumption data, as follows:

- DataPlaneRequests
- MongoRequests
- QueryRuntimeStatistics
- PartitionKeyStatistics
- PartitionKeyRUConsumption
- ControlPlaneRequests
- CassandraRequests
- GremlinRequests
- Requests

In this practice exercise, you'll define diagnostic settings for your Cosmos DB. Follow these steps:

1. Open *https://portal.azure.com* and use your credentials to log in.
2. Navigate to **Cosmos DB**.
3. In the **Monitoring** group in the left panel, click **Diagnostic Settings**.
4. On the **Diagnostic Settings** page, click the **Add Diagnostic Setting** link.
5. In the **Diagnostic Setting Name** box, type a name for the diagnostic setting. For this practice exercise, type **ADSCosmosDBLogs**.
6. Under **Destination Details**, select **Send to Log Analytics**.

7. Enable the following options:
    - **DataPlaneRequests**
    - **QueryRuntimeStatistics**
    - **PartitionKeyStatistics**
    - **PartitionKeyRUConsumption**
    - **Requests** (from the metrics group)
8. Click the **Save** button in the toolbar along the top of the page.
9. To add entries to the logs, open **Data Explorer** and execute the following scripts:

```
SELECT * FROM c
WHERE c.Code = "Country"
SELECT * FROM c
WHERE c.Code = "Data"
```

Audit logs

To manage your resource logs, click **Logs** under **Monitoring** in the left panel of any resource in Azure Portal. A panel with a directory tree opens. This tree contains nodes that represent various logs. In addition, you'll see a work area to the right. (See Figure 4-10.)

FIGURE 4-10 The page where you audit your logs

The logs in the directory tree are as follows:

- **AzureActivity** Contains logs from the Azure platform itself at a subscription or management-group level. Unless something happens at these levels, there will be no entries here.

- **AzureDiagnostics** Contains information about all your resource logs. If you want to look for specifics, like logs for Cosmos DB, they will be filtered by that type of resource.

- **AzureMetrics** Where metrics are copied to logs. These are also filtered by resource type.

If you hover over a node, you will see a description of its contents. (See Figure 4-11.)

FIGURE 4-11 See a description of a node's contents and access a preview of the data here

A **Preview Data** link also appears. Click this link to open a panel with 10 entries from the data. You can scroll through this panel to see the different columns and values the logs display.

At the bottom of the panel is a **See in Query Editor** button. Click this to send the query to the Query Editor (the top pane in the work area to the right of the directory tree), where you can modify it. For example, doing this for the **AzureDiagnostics** node generates the following query:

```
AzureDiagnostics
  | where TimeGenerated > ago(24h)
  | limit 10
```

This query is in Kusto Query Language (KQL), a useful language for performing read-only actions. It uses pipeline redirection, represented with the | character. This character means that the result of the left statement will be input to the right one. The first element in the syntax is the source dataset, followed by filters, limiters, selections, and so on. So, the previous query could be interpreted as, "get AzureDiagnostics items where the TimeGenerated column is greater than the time 24 hours ago, but limit the results to 10 elements."

To obtain the results for the query, click the **Run** button. Several buttons and drop-down lists appear at the top of the pane, which you can use to configure the results. For example, you can select the columns to display or change how results are displayed — say, in chart form (assuming the results are compatible with some form of graphic presentation).

You can also enable the **Group Columns** feature. When you do, a gray row appears, where you can drag one or more columns to group entries. To see how this works, scroll to the right in the list of AzureDiagnostics items obtained from Cosmos DB and locate the **partitionKey_s** column. Then drag and drop the column into the gray drop area to obtain the results grouped by this column.

> **NOTE** This grouping mechanism is for the current view only. If you edit the query — for example, changing the limit value to 50 — and run the query again, the grouping will disappear.

KQL has several statements to sort, filter, group, calculate, and aggregate information. For example, you can easily determine how many events occur by partition key using the following query:

```
AzureDiagnostics
 | summarize event_count = count() by partitionKey_s
```

This query is a good candidate to change the result view to a chart — for example, a donut chart, like the one in Figure 4-12.

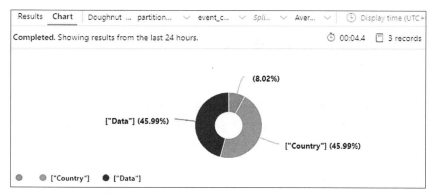

FIGURE 4-12 Donut chart for partition keys

NEED MORE REVIEW? For more about KQL, including syntax and samples, see *https://docs.microsoft.com/en-us/azure/data-explorer/kusto/query/.*

PRACTICE Deploy Azure SQL Analytics

Just as you can analyze logs for a single resource, you can quickly analyze logs for several resources and even the entire subscription. You do this with Azure SQL Analytics. This practice exercise guides you through deploying of Azure SQL Analytics and using it to analyze your subscription.

1. Open *https://portal.azure.com* and use your credentials to log in.

2. In the search box, type **SQL Analytics**, and select the corresponding entry from the drop-down list that appears under **Marketplace**.

3. In the **Create New Solution** page, select the appropriate Azure Log Analytics workspace. Then click the **Create** button.

4. When the deployment is complete, click the **Go to Resource** button.

5. In the **Overview** page, click the **View Summary** link on the **Summary** tile.

   This loads a list of SQL resources, including databases, managed instances, elastic pools, and managed databases.

6. Click a resource — in this case **Databases** — to see a series of database-related analytic charts.

   NOTE You can only select already-deployed resources.

EXERCISE 4-2 Set up a monitor dashboard

In this exercise, you will create a custom dashboard that obtains information from various logs and metrics to enable you to evaluate your resources.

1. Open *https://portal.azure.com* and use your credentials to log in.

2. In the search box, type **Monitor**, and select the corresponding entry from the drop-down list that appears.

   The Overview page opens. It allows easy access to the three major areas to monitor: Metrics, Search Logs, and Alerts.

3. Click **Search Logs** and close the overlaid tab.

   You will see the same page you saw when you clicked the Logs option in the left panel in Cosmos DB. (Refer to Figure 4-10.) Moreover, if clicking the Logs option is the last action you performed, the page will already be filtered by Cosmos DB logs.

4. Click the **Select Scope** link at the top left of the panel.

   A panel opens to the right.

5. Click the **Clear All Selections** button at the bottom of the panel.

6. Select **check box** next to your resource group and click the **Apply** button.

7. In the directory tree to the left, expand the **Azure Cosmos DB** node.

8. Click the **AzureDiagnostics** node and then click the **Preview Data** link.

9. Click the **See in Query Editor** link to send the preview data to the Query Editor.

> **NOTE** AzureDiagnostics is a generic category and may include data from other resources, like Azure SQL Database.

10. Optionally, try different query combinations to practice.

11. Enter the following query text and test it:

```
AzureDiagnostics
  | where TimeGenerated > ago(24h)
  | where  ResourceProvider has("MICROSOFT.DOCUMENTDB")
  | where Category has("PartitionKeyRUConsumption")
  | summarize Units=sum(todecimal(requestCharge_s)) by  collectionName_s
  | render  piechart
```

12. Click the **Pin to Dashboard** button in the toolbar above the Query Editor to add this query to the dashboard you created earlier (ADS My Dashboard).

13. Click the plus sign to the right of the **Query-1** tab to open a new query window.

14. Type the following query:

```
AzureMetrics
  | where TimeGenerated > ago(24h)
  | project ResourceProvider, MetricName,Total,UnitName
  | summarize sum(Total) by MetricName,ResourceProvider,UnitName
  | render barchart
```

15. Review the result. Notice that it is useless because it contains values from so many different counters.

16. Change the `render` option to `table` and review the values returned.

17. Change the query to return something more informative. In this case, run the following query to obtain information about SQL Server usage:

```
AzureMetrics
  | where TimeGenerated > ago(24h)
  | project ResourceProvider, MetricName,Total,UnitName
  | order by MetricName|where ResourceProvider has("MICROSOFT.SQL")
  | where UnitName has("Percent")
  | where Total !=0
  | summarize avg(Total) by MetricName
  | render columnchart
```

18. Repeat step 12 for this query.

> **NEED MORE REVIEW?**   For more information about writing KQL queries, see
> *https://docs.microsoft.com/en-us/azure/data-explorer/kusto/query/sqlcheatsheet.*

Monitor data processing

When you need to analyze data over time, your processes must be reliable and could increase resource consumption. Sooner or later, you will need to fine-tune resource usage, collect information about these resources, and analyze it in sync with other resources. This is the key to better using your data implementation.

Monitor Azure Data Factory pipelines

Azure Data Factory enables you to gather metrics about pipelines, as well as generate logs and alerts to help you manage them.

Metrics

Azure Data Factory pipelines automatically store metrics during execution. You can see these metrics on the Overview page for your Data Factory account in Azure Portal. To access them, click the **Metrics** option in the **Monitoring** group in the left panel. Table 4-8 lists several of these metrics.

TABLE 4-8   Azure Data Factory metrics

Metric Type	Metrics
Pipelines	Failed Pipeline Runs
	Succeeded Pipeline Runs
	Cancelled Pipeline Runs
Activities	Failed Activity Runs
	Succeeded Activity Runs
	Cancelled Activity Runs

Metric Type	Metrics
Triggers	Failed Trigger Runs
	Succeeded Trigger Runs
	Cancelled Trigger Runs
SSIS	Failed SSIS Integration Runtime Start
	Succeeded SSIS Integration Runtime Start
	Cancelled SSIS Integration Runtime Start
	Stuck SSIS Integration Runtime Stop
	Succeeded SSIS Integration Runtime Stop
	Succeeded SSIS Package Execution
	Failed SSIS Package Execution
	Cancelled SSIS Package Execution
Runtime and Engine	Integration Runtime CPU Utilization
	Integration Runtime Available Memory
	Integration Runtime Queue Duration
	Integration Runtime Queue Length
	Integration Runtime Available Node Count
	Maximum Allowed Entities Count
	Maximum Allowed Factory Size (in GBs)
	Total Entities Count
	Total Factory Size (in GBs)

NOTE These metrics are collected by Azure and can be queried from Azure Log Analytics.

You can combine multiple metrics in the same chart. (See Figure 4-13.) However, these should be correlated. For example, it makes no sense to share in the same chart metrics measured in millions with other metrics measured in values between 0.1 and 1.

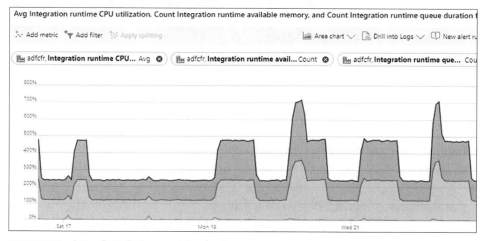

FIGURE 4-13 Azure Data Factory metrics chart

Sometimes, the graph engine will try to correlate data using strange values in an attempt to make the chart reflect the data. For example, in Figure 4-13, the Y axis uses values between 0 and 800 percent because one of the counters is an average while the others are a count. But the chart does properly display the variation in activities between different days.

Logs

In addition to generating metrics, Azure Data Factory can write logs. You can configure these from the **Diagnostic Settings** option in the **Monitoring** group in the left pane of the Azure Data Factory Overview page. Table 4-9 lists the available logs.

TABLE 4-9   Azure Data Factory logs

Log Type	Logs
Logs	ActivityRuns
	PipelineRuns
	TriggerRuns
	SSISPackageEventMessages
	SSISPackageExecutableStatistics
	SSISPackageEventMessageContext
	SSISPackageExecutionComponentPhases
	SSISPackageExecutionDataStatistics
	SSISIntegrationRuntimeLogs
Metrics	AllMetrics

Alerts

You can configure alerts in Azure Data Factory to immediately report issues with metrics or logs, just as you can with several other Azure resources.

Azure Data Factory provides yet another way to monitor logs: from the Azure Data Factory site at *https://adf.azure.com/*. You reach this site by clicking the **Author and Monitor** button in the service's **Overview** page in Azure Portal.

The Overview page of the Azure Data Factory site displays the status of pipelines and activities as donut charts. These identify pipelines and activities that have succeeded or that require attention in percentage form.

In the left panel, the second button corresponds to the Monitor area, where you can see the following information:

- **Runs**   These include pipeline runs and trigger runs.
- **Runtime and Sessions**   These include integration runtimes and a Data Flow Debug option.
- **Notifications**   These include alerts and metrics.

The Runs options list pipelines and triggers that are currently running or have recently run and their status. Each entry has a corresponding check box, which you can select to cancel or debug the pipeline.

To follow the activity of different pipelines graphically, you can change the display from list view to Gannt view. Clicking a bar in Gantt view opens a pop-up panel that shows the pipeline's status and various statistics. (See Figure 4-14.)

> **NOTE** In Gantt view, the bar color represents its status: red for error, blue for in progress, and green for success.

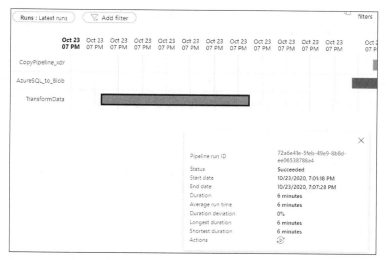

FIGURE 4-14 Azure data pipelines in Gantt view

Monitor Azure Databricks

The Azure Databricks service does not automatically present metrics like other Azure services do. Instead, you must add your own code in the workspace to create a job that will send metrics to Azure Log Analytics.

PRACTICE   Send Azure Databricks metrics to Azure Log Analytics

To send Azure Databricks metrics to Azure Log Analytics, follow these steps:

1. Open *https://portal.azure.com* and use your credentials to log in.
2. Click the **Databricks** tile.
3. In the **Overview** page, click the **Launch Workspace** button to launch the Databricks workspace.
4. Click the **user** icon in the top-right corner of the page and select **User Settings** from the menu that appears.

   You must create a token to reach the workspace from an API without a password.

5. In the **Access Tokens** tab of the **User Settings** dialog box, click the **Generate New Token** button to create a new token.

6. Optionally, clear the **Lifetime** box to make the token permanent.

   Now you can configure your workspace to send metrics to Azure Log Analytics.

7. Back in Azure Portal, open an Azure Cloud Shell and set it to use bash.

8. In Azure Cloud Shell, create a virtual environment.

9. Enter the following code to switch to the new virtual environment:

```
virtualenv -p /usr/bin/python2.7 <Virtual_Environment_Name>
source <Virtual_Environment_Name>/bin/activate
```

10. Use the following code to install the Databricks CLI:

```
pip install databricks-cli
```

11. Use the following code to authenticate with the token you created previously:

```
databricks configure --token
```

> **NOTE** You will be asked to enter the URL for the workspace and the token. This will be something like *https://adb-<number_id>.azuredatabricks.net*.

12. Prepare and install the Spark monitoring library.

> **NOTE** You can find the source code and procedures to do this in Docker or using Maven here: *https://github.com/mspnp/spark-monitoring*.

13. Employ the **UserMetricsSystem** library created in the previous step to define your counters.

   For a sample in Scala, see *https://github.com/mspnp/spark-monitoring/blob/master/sample/spark-sample-job/src/main/scala/com/microsoft/pnp/samplejob/StreamingQueryListenerSampleJob.scala*.

   Alternatively, you can define your counters for log4j using the following code:

```
log4j.appender.A1=com.microsoft.pnp.logging.loganalytics.LogAnalyticsAppender
log4j.appender.A1.layout=com.microsoft.pnp.logging.JSONLayout
log4j.appender.A1.layout.LocationInfo=false
log4j.additivity.com.microsoft.pnp.samplejob=false
log4j.logger.com.microsoft.pnp.samplejob=INFO, A1
```

   The information will appear in Azure Log Analytics here:

   - SparkLoggingEvent_CL for logs
   - SparkMetric_CL for metrics

NEED MORE REVIEW? For more information about defining and managing logs in Azure Databricks, see *https://docs.microsoft.com/en-us/azure/architecture/databricks-monitoring/application-logs*.

NEED MORE REVIEW? There is an automatic log-collection and delivery feature in preview right now. To read more about it, see *https://docs.microsoft.com/en-us/azure/databricks/administration-guide/account-settings/azure-diagnostic-logs*.

Monitor Azure Stream Analytics

When using Azure Stream Analytics, you can monitor jobs directly in the service's Overview page. This page displays the most important information — input and output events, runtime errors, and resource utilization — in chart form. (See Figure 4-15.)

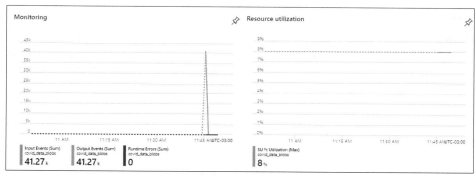

FIGURE 4-15 Charts on the Azure Stream Analytics Overview page

Metrics

As with most Azure resources, with Azure Stream Analytics, you can select your own combination of metrics to evaluate, add them to charts, and pin them to your dashboard. Table 4-10 lists the metrics available for Azure Stream Analytics.

TABLE 4-10 Azure Stream Analytics metrics

SU % Utilization	Function Requests
Input Events	Failed Function Requests
Input Event Bytes	Function Events
Late Input Events	Input Deserialization Errors
Output Events	Early Input Events
Data Conversion Errors	Watermark Delay
Runtime Errors	Backlogged Input Events
Out of Order Events	Input Sources Received

Logs

Like most other Azure resources, Azure Stream Analytics can send log information to be used by Azure Log Analytics and other services. Table 4-11 lists the log sources available for Azure Stream Analytics.

TABLE 4-11   Azure Stream Analytics log sources

Type	Log Source
Logs	Execution
	Authoring
Metrics	All Metrics

> **NOTE**   Procedures for querying and charting results in Azure Stream Analytics are identical to those for storage resources.

Monitor Azure Synapse Analytics

As with other services, you can use the Monitor and Diagnostic Settings links in the Azure Synapse Analytics Overview page to retrieve information about it. There are two groups of metrics for Azure Synapse Analytics. (See Table 4-12.) You use these metrics in exactly the same way as with other Azure services — including adding them to your dashboards if you wish.

TABLE 4-12   Azure Synapse Analytics metrics

Type	Metrics
For SQL pools	Data Processed
	Login Attempts
	Requests Ended
Integration	Activity Runs Ended
	Pipeline Runs Ended
	Trigger Runs Ended

Monitor SQL pools using DMVs

Optimizing Azure Synapse Analytics is about properly managing stored data and data processing. Certain configurations can enhance this. For example, if you enable `auto_create_stats` in your SQL pools, the platform will better estimate the best way to perform each task. To see if this setting is enabled in your database pool, run the following script:

```
SELECT
      [is_auto_create_stats_on]
   FROM
      [sys].[databases]
   WHERE [name] = '<SQL Pool Name>';
```

```
SELECT
        [is_auto_update_stats_on]
    FROM
        [sys].[databases]
    WHERE [name] = '< SQL Pool Name'>';
```

However, this is not supported in this type of database.

As with Azure SQL Database, you can use DMVs to evaluate how processes in Azure Synapse Analytics are working. Remember, though, that the user querying the DMVs must have proper permissions. You can assign these with the following script:

```
GRANT VIEW
DATABASE STATE TO [<User_Name>];
```

You can analyze the behavior and performance of your queries by querying the sys.dm_pdw_exec_requests DMV. For example, the following query returns the 10 most time-consuming calls:

```
SELECT TOP (10)
        [command],
        SUM([total_Elapsed_time]) AS [Sum Of Time],
        COUNT(*) AS [Qty]
    FROM
        [sys].[dm_pdw_exec_requests]
    GROUP BY
            [command]
ORDER BY
        SUM([total_Elapsed_time]) DESC;
```

If you want to evaluate the execution plan for the most time-consuming queries, you can use a script like this:

```
SELECT
        *
    FROM
        [sys].[dm_pdw_request_steps]
    WHERE [request_id] IN
    (
        SELECT TOP (10)
                [request_id]
            FROM
                [sys].[dm_pdw_exec_requests]
        ORDER BY
                [total_Elapsed_time] DESC
    )
```

```
ORDER BY
        [request_id],
        [step_index];
```

Other useful DMVs include the following:

- `sys.dm_exec_sessions`
- `sys.dm_pdw_dms_workers`
- `sys.dm_pdw_sql_requests`
- `sys.dm_pdw_waits`

When working with lots of information and jobs, it can become very difficult to find infor-mation about a specific query. You can use the LABEL option for those queries you suspect could be problematic to filter the DMVs by the label, as in the following query:

```
SELECT
        [s].[ProductKey],
        [EnglishProductName] AS [OrderDateKey],
        [DueDateKey],
        [ShipDateKey],
        [ResellerKey],
        [EmployeeKey],
        [PromotionKey],
        [CurrencyKey],
        [SalesTerritoryKey],
        [SalesOrderNumber],
        [SalesOrderLineNumber],
        [RevisionNumber],
        [OrderQuantity],
        [UnitPrice],
        [ExtendedAmount],
        [UnitPriceDiscountPct],
        [DiscountAmount],
        [ProductStandardCost],
        [TotalProductCost],
        [SalesAmount],
        [TaxAmt],
        [Freight],
        [CarrierTrackingNumber],
        [CustomerPONumber]
FROM
        [dbo].[FactResellerSales] AS [s]
            INNER JOIN
                [dimproduct] AS [p]
            ON [s].[ProductKey] = [p].[ProductKey]
    OPTION (LABEL='FPROD');
```

This query returns information about the execution of the query by using the label:

```
SELECT
      [total_elapsed_time],
      [row_Count],
      [operation_type]
FROM
      [sys].[dm_pdw_request_steps]
WHERE [request_id] IN
  (
     SELECT TOP (10)
           [request_id]
       FROM
            [sys].[dm_pdw_exec_requests] AS [r]
        WHERE [r].[label] = 'FPROD'
        ORDER BY
                [total_Elapsed_time] DESC
   )
ORDER BY
        [request_id],
        [step_index];
```

Another interesting DMV is sys.dm_pdw_dms_workers. This lets you analyze how data moves between different workers. With this DMV, you can investigate the bytes processed, rows processed, total elapsed time, and other information related to the data movements involved in a query, filtering by request_id.

Monitor Spark pools

Spark pools have a different set of metrics:

- Active Apache Spark Applications
- Ended Apache Spark Applications
- Memory Allocated
- vCores Allocated

To research these, you must navigate to each pool. Here's how:

1. Open *https://portal.azure.com* and use your credentials to log in.
2. Navigate to your Azure Synapse Analytics workspace.
3. In the left panel of the **Overview** page, under **Analytic Pools**, click **Apache Spark Pools**.
4. Click the desired pool name.
5. In the **Monitoring** group in the left panel, click **Metrics**.

Spark pool logs

As with other services, Spark pools have a set of log sources. This includes the BigData-PoolAppsEnded log. You include it in your log analysis the same way as you do logs for other Azure services.

PRACTICE Add the BigDataPoolAppsEnded log

To add the BigDataPoolAppsEnded log, follow these steps:

1. Follow steps 1 through 4 in the preceding section to view the options for a particular pool.
2. Click the **Diagnostic Settings** link in the **Monitoring** group in the left panel of the Spark pool's page.
3. In the **Settings** page, click the **Add Diagnostic Setting** link.
4. Select **BigDataPoolAppsEnded**.
5. Select the **Send to Log Analytics Workspace** and **Archive to a Storage Account and/or Stream to an Event Hub** check boxes.
6. Add the appropriate values for each destination.

> **NOTE** It will take some time for information to populate the destinations to query the log.

As its name suggests, the BigDataPoolAppsEnded log reports activities that have already ended, including the following information:

- TenantId
- TimeGenerated [UTC]
- OperationName
- Category
- CorrelationId
- Type
- _ResourceId
- Properties

With regard to Properties, this includes the following information:

- incomingCorrelationId
- applicationName
- schedulerState
- applicationId
- pluginState
- submitterId

- submitTime
- startTime
- endTime

If you query the log, it will contain one entry for each time a Spark session has ended and use the endTime and startTime properties to get the periods of usage.

Configure Azure Monitor alerts

Using Azure Monitor to manage data-processing metrics processing is identical to using it to monitor data storage, as discussed earlier.

You can query monitor data externally with the same tools you use for other Azure tasks, like PowerShell, Azure CLI, the REST API, and .NET libraries. The following code sample, available in the companion content for this chapter in the Get Metric Values.ps1 file, is a template to query values from a metric counter:

```
$SubscriptionId="<Your_SubscriptionId>"
$resourceId="subscriptions/<Your_SubscriptionId>/resourceGroups/<Your_Resource_Group>/
providers/<desired_Provider>/<Resource_Name>"
$MetricName='<Metric Name>'

#Load the Azure Snap-in
Import-Module -Name Az
#Connect to your Azure Account
Connect-AzAccount
$context=Get-AzContext
Get-AzSubscription
Set-AzContext -SubscriptionId $SubscriptionId
Clear-Host
$metrics=Get-AzMetric `
    -ResourceId $resourceId `
    -MetricName $MetricName `
    -TimeGrain 00:01:00   `
    -ResultType Data |`
        select Timeseries
foreach($metric in $metrics.Timeseries)
{
    $metric.Data|Format-Table
}
```

> **NOTE** You can obtain the **ResourceId** value from your resource's Overview page. Simply copy the URL from the slash before *subscriptions* to the slash before *Overview*.

Lately, Microsoft has proposed using Grafana as part of a complex environment to present information graphically. Grafana is an independent open source product. To implement Grafana, you must deploy it in a VM in your resource group. Then, configure it to authenticate itself properly with Azure Active Directory. This will enable it to connect to Azure Log Analytics, where it will retrieve all the information.

Audit with Azure Log Analytics

For each resource in the "Monitor data processing" section, there are details about how to configure logs to send information to Azure Log Analytics. These configurations relate to your own processes. But there are other logs you need to configure if you want to optimize resource utilization.

Each resource has an **Activity Log** option in the left panel. You can click this button to open an Activity Log page, which contains a list of resource activities. These activities relate to the service itself rather than its processes.

To create or manage diagnostic settings for a resource, click the **Diagnostic Settings** button in the toolbar at the top of the resource's Activity Log page. For example, in the Activity Log page for Azure Data Factory, you can choose from among the following diagnostic settings:

- Administrative
- Security
- Service Health
- Alert
- Recommendation
- Policy
- Auto Scale
- Resource Health

Use Azure Log Analytics notebooks

In your day-to-day work, you will add more deployments and resources. Sooner or later, you will want to monitor and follow up on each of them.

Having several people working as a team enables you to provide support for longer durations, covering time periods by alternating people. However, this requires good documentation and regular follow-ups that involve various step-by-step procedures, all of which must be well documented. Azure Log Analytics notebooks can help with this.

Azure Log Analytics notebooks enable you to gather information to review and explain what procedures to use to follow up on any log data you receive — for example, when it reveals some type of risk. The structure of these notebooks is identical to that of the Jupyter Notebooks you used in Chapter 1 for Databricks, but with metrics and charts from logs.

You can add the following types of content to Azure Log Analytics notebooks:

- **Text**   This is for your own content. You can use the following text styles:
  - Plain
  - Info
  - Upsell
  - Success
  - Warning
  - Error

  Also, except with the Plain style, you can add icons to the text to flag or identify it.

- **Parameters**   You can define several parameters, which can be entered by whoever is using the notebook, and whose values can be made available to other content in the notebook, like metrics or queries. A Parameter Editor enables you to use different data types and obtain data from different sources — for example, common elements from Azure, like resource types. You can also use the Parameter Editor to perform queries against your data, obtain the names of resources in logs or metrics, and so on. Finally, you can build a custom list in JSON.

- **Links and tabs**   You can use these to create shortcuts to different parts of the Azure Portal and to other content. These can be displayed as lists, lists of links, tabs, and so on.

- **Queries**   You can query information in Azure Log Analytics from any resource or type, using parameters you define. For example, the following query sample uses a ResourceProvider parameter to filter information by resource and display it in a chart.

```
AzureMetrics
 | where TimeGenerated > ago(24h)
   | where ResourceProvider has '{ResourceProvider}'
    | project MetricName,Maximum,Minimum
      | render areachart
```

- **Metrics**   You can add one or more metrics to a chart, obtaining the chart information directly from the metrics. The metrics chart editor is the same one you've already seen for other resources but embedded into the notebook.

- **Groups**   Think of a group as a notebook inside a notebook. When you create a group, it can contain any of the other elements described previously, plus other groups hierarchically.

> **NOTE**   You can create more than one notebook and combine them.

EXERCISE 4-3   Create an Azure Log Analytics notebook

In this exercise, you will create an Azure Log Analytics notebook to summarize the most important information from your resources.

1. Open *https://portal.azure.com* and use your credentials to log in.

2. In the search box, type **Log Analytics workspaces**, and select the corresponding entry from the drop-down list that appears.

3. Click your workspace.

4. In the left panel, under **General**, click **Workbooks**.

> **NOTE**   Some samples appear, which you can examine if you want.

5. Click the **Empty** option under **Quick Start** to create a new empty workbook.

6. To create a name for your workbook, click the **down arrow** to the right of the **Add** button and select **Text**.

7. Enter the following text:

```
# ADS Demo Workbook
## This workbook will contain a Parameter selector to filter the information
displayed in it.
### The parameter will be used in a log query.
```

> **NOTE**   You can use HTML notation, but markdown notation is preferred.

> **NOTE**   The # symbols define the heading level.

8. Click the **Done Editing** button.

9. Click the **down arrow** to the right of the **Add** button and select **Parameters**.

10. Click the **Add Parameter** button.

   A panel opens to the right.

11. Configure the following parameter settings:

   - **Name** Type **Period** in this box. This will be the parameter's internal name.
   - **Display Name** Again, type **Period**. This will be the parameter's display name.
   - **Parameter Type** Choose **Time Range Picker**.
   - **Required** Select this check box.
   - **Explanation** Select the period to evaluate.

12. Click **Save**.

13. Repeat steps 9 and 10 to add another parameter — this time, one that filters by resource.

14. Configure the following parameter settings:

   - **Name** Type **Resource** in this box.
   - **Display Name** Again, type **Resource**.
   - **Parameter Type** Choose **Drop Down**.
   - **Required** Select this check box.
   - **Explanation** Select the resource to display.
   - **Get Data From** Choose **Query**. Then enter the following query to obtain the available resources in your logs:

```
AzureMetrics
    | project ResourceProvider
    | distinct ResourceProvider
```

15. Click **Save**.

16. Click the **Done Editing** button.

17. Click the **down arrow** to the right of the **Add** button and select **Query**.

18. Type the following query:

```
AzureMetrics
    | where ResourceProvider has "{Resource}"
    | project MetricName,Maximum,Minimum
    | render areachart
```

19. Open the **Time Range** drop-down list and select **Period**.

   This appears at the bottom of the list and is the Period parameter you just created.

20. Click the **Done Editing** button.

21. Test your development by changing values in the **Parameters** section to see different query results.

22. Click the **Save** button at the top of the workbook.

23. Name the workbook **ADS Workbook**.

Optimize Azure data solutions

Sooner or later, handling large volumes of data could generate storage or management issues. Moreover, a data-analysis application might be stressed not only by the volume of data, but also by the complexity of several services interacting and competing for resources. This section analyzes how to resolve these and other critical issues.

Troubleshoot data-partitioning bottlenecks

Implementing a data-analysis application involves managing very large data stores. To achieve this, and optimize the process at the same time, you must define partitions for the data. This will improve performance and scalability while also reducing contention.

In some cases, managing partitions could be the responsibility of the platform. In others, it will be your responsibility. Of course, even if your partitions are managed by the platform, you will still need to define them.

Data can be partitioned in three different ways:

- Horizontally
- Vertically
- Functionally

Horizontal partitioning

Also called *sharding*, horizontal partitioning splits data into segments in different data stores (shards). All the shards in a partition must have the same schema. In addition, you must define a sharding key before implementation. Note that the sharding key can be difficult to change afterward; in most cases, changing the sharding key requires rebuilding the entire storage.

When defining the sharding key, often called a *partition key*, consider how it will be used. Sometimes you will need to retrieve information by specific items using a lookup pattern. For these cases, you could segment the information by, say, region or group, knowing you must always search by this element. Or you might use a range pattern if you know that you usually get information in sets, like when you query by periods of time. Finally, a hash pattern involves using a hash based on keys and data size. This prevents the system from overloading one partition compared to the others because some sources generate larger datasets or more data.

Defining horizontal partitions is not about distributing data equally. It's about distributing it by frequency of use. You can have a very large partition but with low access — for example, with historical data, which is not required in every query.

> **NOTE** Ideally, the partition key will be immutable, to prevent the movement of data between shards.

Vertical partitioning

With vertical partitioning, partitions are based on how frequently the attributes of the data are used. Different properties of the same entity could be used more often than others. Analyzing this could guide you on how to split the information into different partitions. Given this, the schema will not be equal between partitions, because in fact they will contain different information. This partition strategy is used frequently in HBase and Cassandra and is sometimes implemented in Azure SQL Database using column stores.

Functional partitioning

Functional partitioning partitions your data based on how you use it. For example, you might have partitions based on operational areas like Sales, Human Resources, Production, and so on. Or you might have one set of partitions for read-only data and another for frequent write-read access. In this case, the distribution will be managed by your implementation instead of by the platform, which means more control — and work — from your side.

Partitioning considerations

Sometimes, the usage scenario changes how you implement partitions. That is, you implement different partitions when you need scalability versus query performance, and so on. Following are several tips for different scenarios.

Partitions for scalability

This is a list of important points to consider when partitioning for scalability:

- The size of the result sets usually retrieved, which implies end-to-end latency
- The frequency of the queries
- The complexity of the queries, which imply server-side latency
- Ensuring adequate resources for each partition
- Monitoring the workload to redistribute partitions when appropriate

Partitions for query performance

There are two important considerations when partitioning for query performance:

- Smaller result sets
- Parallelism

When partitioning for query performance, reviewing the business requirements will usually guide you to the best approach. Here are some steps you might take:

- Research to figure out which queries are critical and need to be executed more quickly.
- Evaluate and monitor metrics to identify slower queries.
- Identify repetitive queries. Even if they can be executed quickly, repetitive queries may degrade performance if they are executed often. A good way to avoid this could be to cache the results and decide to partition in a different way.
- Partition horizontally and define a good partition key to make it easier for the platform to find the required data and avoid searching records one by one on every partition.

Partitions for availability

When partitioning for availability, you must establish partitions to protect data and ensure that recovery processes are both quick and reliable. These are the important points to keep in mind:

- Keep the most important data in a few partitions and back it up frequently.
- Establish redundancy policies for the partitions that contain the most important data.
- Manage and monitor partitions separately to keep more control over critical data.
- When possible, manage geographical partitions to perform maintenance during out-of-work periods by region.

Partitions and application design

Having data in partitions affects how the application works with them. This could be difficult when implemented on a pre-existing system because the application must be preserved during partition migration.

The data partition is not only about huge volumes of data. Heavily used applications could benefit from using partitions to distribute the workload of simultaneous calls. For these kinds of applications, consider the following tips:

- Avoid cross-partition operations when possible. When parallel queries are not serialized, they can be performed and their results merged at the application level. This might be impossible, however, if some queries depend on the results of previous queries.
- Keep closely related data in the same partition to reduce query complexity.
- Replicate referential data — for example, country codes, product categories, and so on — between partitions to avoid cross-partition access. At the same time, depending on the volume of this data, consider loading it in memory at the application start.
- Denormalize data to avoid referencing cross-partition queries. Usually, data partitioning is not a good strategy for intense OLTP processes. When you need to use partitions for these kinds of solutions, consider replication processes to maintain copies of the most-referenced data in all the partitions.

- Work with lower consistency levels when possible. Using eventual consistency might not be possible in all cases, but most distributed operations could rely on it without affecting functionality.

- Avoid distributed transactions. It is better to have an atomic transaction in just one partition and use background replication for data that needs to be available in other partitions.

Partition Azure SQL Database

A single Azure SQL Database might not be adequate to store all your data. In this case, you can implement an *elastic pool*. Elastic pools split data into different databases called *shardlets*, identified by shardlet keys.

Each shardlet stores metadata to maintain information about the shardlet. A central shard-map database is used to coordinate different updates and queries against the data. This enables you to maintain segmented information that acts exactly like it would if it were partitioned on other storage platforms.

The shard-map database could contain shard-key ranges and/or lists of individual shardlet keys. For example, health information stored in an elastic pool could be defined by a *shardlet* key that uses the country code. Then, information related to diseases and treatments could be stored in different databases based on that code. Figure 4-16 shows a schema of an elastic pool.

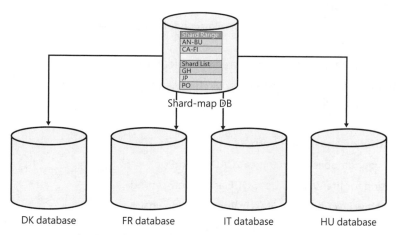

FIGURE 4-16 Elastic pools Azure SQL Database partition schema

Elastic pools enable an API to create new shards when required and to simplify data transfer from applications into the elastic pools.

NEED MORE REVIEW?   There are libraries for managing elastic pools for Java and .NET. Read about their implementation and download them here: *https://docs.microsoft.com/en-us/ azure/azure-sql/database/elastic-database-client-library.*

Shard keys can be of the following data types:

- Integer
- Long
- Guid
- Byte[]
- Datetime
- Timespan
- Datetimeoffset

NEED MORE REVIEW?   You use code to manage your shard-map database. For implementation details and code samples, see *https://docs.microsoft.com/en-us/azure/azure-sql/ database/elastic-scale-get-started.*

PRACTICE   Create an elastic pool

After a shard key has been defined, you can create an elastic pool. This involves adding a new Azure SQL Database. Follow these steps:

1. Open *https://portal.azure.com* and use your credentials to log in.
2. In the search box, type **SQL Databases**, and select the corresponding entry that appears in the drop-down list.
3. Click the **Add** button.
4. Select your subscription and resource group.
5. Type a name for your database — in this case, ADSDBforPool.
6. Click the **Yes** option for the Want to Use SQL Elastic Pool? setting.
7. Click the **Create New** link under the **Elastic Pool** drop-down list.
8. Type a name for the elastic pool — in this case, ADS_EPool.
9. Under **Compute + Storage**, click the **Configure Elastic Pool** link.
10. Click the **Change Configuration** link and select the desired hardware configuration from the list that appears— for this example, Gen5.

> **NOTE**   You can also change the vCores quantity and maximum data size. This changes the cost, however.

11. Leave the **Zone Redundancy** setting disabled and click the **Apply** button.

12. Click the **Additional Settings** tab and select **Sample** as data source.

13. Click the **Review + Create** tab. Then click the **Create** button.

14. When this deployment is complete, repeat these steps to create another new SQL Database, but make the following selections:
    - Type **ADSDBforPool2** for the database name.
    - Select the previously created elastic pool.

15. When the second deployment is complete, switch to that resource and click the name of the elastic pool.

    The elastic pool's Overview page opens.

16. Click the **Configure** option in the left panel or the **Configure** button along the top of the page to review the configuration and both databases under the pool.

> **NOTE** There is a button to add more databases on the same page.

> **NOTE** With high data volume and frequent inserts and updates, you will need to rearrange your shards to get better performance. There is a tool for this, called the split-merge tool. To obtain this tool and learn how to configure it, see *https://docs.microsoft.com/en-us/azure/azure-sql/database/elastic-scale-overview-split-and-merge*.

Partition Azure Blob storage

Things are different when partitioning Azure Blob storage. A *blob* is a unit of information in Azure Blob storage, even if it is built by several blocks. The information in an Azure Blob storage account can be replicated, but always as entire blobs. There is no block partitioning.

The partition key in Azure Blob storage is the unique ID of the blob, which is a combination of the account name, the container name, and the blob ID, as in the following example: adss-toragedatalakeADS.blob.core.windows.net/globaldata/Daily/ABW.

Azure storage internally manages data partitions based on ranges of partition keys. For better partition management, consider using a three-digit hash for your naming convention; this optimizes the search by ranges. In a similar way, when using timestamps, use the seconds as the first part of the blob ID, as in *ssyyyymmdd*. This enables the platform to spread the data in six ranges of 10 seconds, up to 60 ranges, to make it easier to obtain segmented information.

Storage accounts tend to keep all blobs for a container in the same partition unless they overload the server capacity. In that case, the system will rearrange the blobs by their partition keys.

Partition Cosmos DB

Cosmos DB internally implements physical partitions based on data size and provisioned throughput. This is something you cannot change. However, the platform stores your data in partitions based on a hash of your logical partitions.

The throughput provisioned is distributed evenly between the partitions, and there is some relationship between your logical partition key and physical partitions. Choosing a good partition key, which divides your data in segments of similar size, is critical to optimizing data distribution.

Data distribution is important not only for size, but for usage as well. If a set of data is more frequently used, and it is in the same partition, then that partition will experience more throughput than the others. This can increase costs for that partition, while other partitions use none of their assigned throughput.

You can use Azure Portal to monitor your Cosmos DB partitions. To view your partitions, click the **Storage** tab and select a database and container. The partitions will be displayed in chart form; select one to view more information about it, such as which ones are logical partitions.

As you can see in Figure 4-17, selecting one or more partitions activates an **Open Data Explorer** button, which enables you to look directly at the data to evaluate partition-key usage. If you click the button, the page changes to an Explorer page by sending a filtered query by the partition keys, like the following:

```
SELECT * FROM c WHERE c.Code = "Data" OR c.Code = "Country"
```

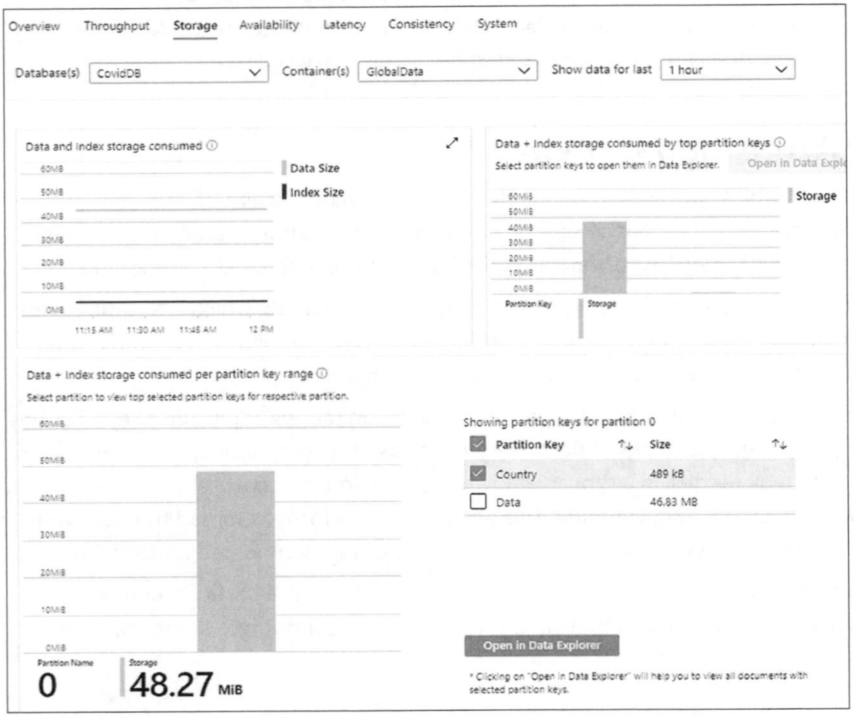

FIGURE 4-17   Partition distribution in Cosmos DB

Optimize Azure Data Lake Storage Gen2

Azure Data Lake Storage (ADLS) Gen2 is an implementation of Azure Blob storage with predefined high throughput to support intensive reads and writes. With the hierarchical namespace enabled, it is used as a data lake for analysis processes.

ADLS is designed to perform as many parallel tasks as needed and can manage very large amounts of data. In the rare case that you need more throughput, you can ask Azure Support for it, but this rarely occurs. Most problems will probably relate more to source or destination issues.

If you need to upload huge volumes of data and repeat this task over time, the data source could become a problem in three key ways:

- **Source hardware and platform** A data lake is designed to store information very quickly. But if the reading process takes time, nothing can be optimized at the destination. Reading from HDD instead of SSD, or slow processing on the source side, will create bottlenecks. A possible solution might be to have a VM receive the data from different sources and send it to the data lake. Or you might have this VM act as a central requester from several different data sources, with enough power to process and send the data to storage more quickly. For these scenarios, you should use Dv4 or Dsv4-series VMs.

> **NEED MORE REVIEW?** For a complete list of VM sizes, see *https://docs.microsoft.com/en-us/azure/virtual-machines/sizes-general*.

- **Network connectivity** Data transfer could be another bottleneck. If you need to transfer from on-premises to a data lake, you can use Direct Connection and Azure ExpressRoute. If you decide to use a VM, you will obtain faster connections with VMs deployed in the same region as the storage — either in the same data center or one that is nearby.

- **Parallelization** With good connectivity, processes sending data to the data lake will take advantage of the destination performance by implementing parallelism. Depending on the scenario, different tools could be used, from custom code to platform tools like Azure Data Factory or Apache DistCp. Each of these requires different configuration procedures and implementation to increase parallelism. For example, in Azure Data Factory, you can enhance performance by using parallel copies or implementing concurrent jobs.

> **NOTE** Choosing the right tool to upload data is not only about process but also about platform. Having different data sources with different volumes could require several tools to perform this task. For examples of different scenarios and information about selecting the best tool, see *https://docs.microsoft.com/en-us/azure/storage/blobs/data-lake-storage-data-scenarios*.

Optimize Azure Stream Analytics

Azure Stream Analytics can enhance performance by managing partitions, depending on the partitions defined for the input and output stores and the streaming processes used. The ideal scenario is when the partition from the source matches a partition at the destination, and at the same time, an instance of the query job matches them. This is called an *embarrassingly parallel job.*

To achieve this level of parallelism, three basic conditions must be met:

- **Source**   The data source must have a defined partition key. This allows the stream process to obtain the data in parallel segments, one per partition. Sources like Event Hubs, Cosmos DB, and Azure tables have defined partition keys. When using Azure Blob storage, folders are used as partition keys. When the hierarchical namespace is enabled, the hierarchy automatically acts as the partition key.

- **Process/query**   The query must match the partition segmentation — for example, the query filter, order, or group information — by using the partition key values. The execution splits in parallel using the partition and reducing the execution time. Older implementations of streams (up to version 1.1) require the use of a PARTITION BY statement in the query to manage the partitions.

- **Destination**   Partitions for destinations are established in the same way as for the source. Having the process match the destination partition allows for parallel execution.

Having the same quantity of partitions in source, process, and destination is the optimal scenario. This level of optimization affects the number of streaming units (SUs) consumed by the process. It is important to estimate and configure SUs to avoid stream job failures. If a job overloads on SUs, it will fail. The standard estimation for a job without partitions is 6 SUs per job. This value can be used as a starting point. So, consider 6 SUs per partition when you estimate the usage. It is a best practice to configure more SUs than necessary to reduce the risk of a failure.

Two metrics, Watermark Delay and Backlogged Events, can be used to analyze whether your SUs are about to be exhausted. It is especially important to monitor and evaluate these metrics when your output target is an Azure SQL Database or a Cosmos DB, because these platforms use different management systems to deal with partitioning. For example, Azure SQL Database requires the Inherit Partitioning setting to be enabled to perform parallel writings.

> ***NEED MORE REVIEW?***   For more information about embarrassingly parallel jobs in Azure Stream Analytics, including query samples and optimization patterns, see *https://docs. microsoft.com/en-us/azure/stream-analytics/stream-analytics-parallelization# embarrassingly-parallel-jobs.*

Optimize Azure Synapse Analytics

When a complex query is about to be executed, the pool moves data between nodes to process it. Examples of complex queries are those that include semi-additive or non-additive measures, like queries with distinct counts or joins by columns that are not distribution columns.

> **NOTE** Large tables, like fact tables, are good candidates for complex query analysis. These queries will be processed in segments to accelerate the process, splitting the queries between nodes in the pool.

Because new tables are usually created with a round-robin distribution, data must move more frequently. This is the reason for the creation of fact tables with hash distributions, as in the following statement:

```
. CREATE TABLE [FactFeedback]
( [ProductKey]       INT NOT NULL,
  Year               INT NOT NULL,
  Month              INT NOT NULL,
  Day                INT NOT NULL,
  [SatisfactionLevel] SMALLINT NOT NULL
)
WITH
(
    DISTRIBUTION = HASH (productKey),
    CLUSTERED COLUMNSTORE INDEX
)
```

When defining a hash distribution, you must think about the query result you want to obtain. You must also consider these important points:

- **Account for the number of entries by hash key** In the previous code, hashing by product key seems like a good idea, but if one of the products has more sales than the others, it will probably overload some nodes.

- **Avoid using nullable values** A null value might make the hashed part bigger than the others.

- **Avoid using columns with default values** This is for the same reason you should avoid using nullable values.

- **Avoid using columns with low cardinality** For example, gender is not a good choice for a hash distribution.

- **Avoid over-partitioning** Azure Synapse Analytics uses 60 databases to distribute work. If you partition to 100 slices, your data will be divided into 6,000 pieces.

This distribution could be applied in certain cases to dimension tables as well. During your day-to-day work, you will monitor your Azure Synapse Analytics processes and detect whether there are any queries slowing down the process. This way, you can reorganize some of the dimension tables that can benefit from using a distribution hash.

Replication strategy

If you have dimension tables in a star schema, certain queries will likely force the replication of the entire table in a round-robin fashion, like the one used by Azure Synapse Analytics. So, if queries can be distributed to up to 60 nodes, there could be 60 replication procedures between nodes. And that could occur for every query performed!

When you know some tables will be used frequently, it is better to force the replication of those tables in the creation process by using a REPLICATION statement, as in the following code:

```
CREATE TABLE [<schema>].[<Table_name>]
<Columns_Definitions>
WITH
(
   DISTRIBUTION = REPLICATE,
   CLUSTERED COLUMNSTORE INDEX
)
```

This way, the data is replicated once during each update procedure instead of each time a query needs to be processed.

Optimize Spark pools

There are different ways to optimize Spark pools depending on what kind of processes you need to perform. For example, you can configure caching or data skewing for the entire pool to improve performance. However, the most important optimization techniques pertain to data management, memory management, and cluster management.

DATA MANAGEMENT

With data management, you can do the following:

- **Change where data is stored** This might be Azure Blob storage, ADLS Gen 2, or local HDFS storage for HDInsight. The fastest is local storage, but the most reliable and durable is ADLS, because local storage will be lost if members of the cluster shut down.
- **Use a different data abstraction** Spark was originally implemented using Resilient Distributed Datasets (RDDs), which can be partitioned to process data in different nodes. A newer format is the dataset, which expands RDDs for use with query engines.

An expansion of the dataset is the data frame, which exposes the dataset simply, like a table. RDD is not the best choice unless you must keep older implementation codes. Between datasets and data frames, the former is easier for development and is checked at compile time but has more overhead due to serialization and garbage collection. The latter has memory optimization features but is checked at runtime. Consider using data frames when you don't have a huge amount of data, but you need to process it quickly, and use datasets for higher volumes.

MEMORY MANAGEMENT

You can enhance memory management by doing the following:

- Configuring small data partitions, which can be easily distributed in the cluster.
- Being mindful of the data types used. Avoid data types that require more space to process, like large strings.
- Using other serializers during data processing. By default, Spark uses the JSON format for serialization, which is the correct one for storing data in data lakes and other external stores. However, when large amounts of data must be distributed between nodes and processed within them, you can probably use other serializers, like Kryo.

> **NEED MORE REVIEW?** Kryo is not the default and requires a bit of configuration. However, this might be worthwhile if you frequently need to process large and complex datasets. You can find the entire Kryo specification code and information about its usage here: *https://github.com/EsotericSoftware/kryo*.

CLUSTER MANAGEMENT

Monitoring the Spark pool enables you to see how its clusters are working. You can then fine-tune the configuration for the following:

- Number of executors (the quantity of nodes running in parallel)
- Number of vCores assigned to each node
- Memory amount for each node

To change your Spark pool configuration, navigate to the pool's Overview page in Azure Portal. Then click the **Auto-Scale Settings** button above the **Essentials** area. This opens a panel to the right, where you can configure the following settings:

- **Auto Scale** You can enable or disable this.
- **Node Size** Your options here are as follows:
  - **Small** Four vCores and 32 GB
  - **Medium** Eight vCores and 64 GB
  - **Large** 16 vCores and 128 GB
- **Number of Nodes** This can be between 3 and 10.

Manage the data life cycle

After you implement your data environment and have all your processes working, your data will grow and use more resources as time passes. When it does, your old data will usually become less important. In other words, it will advance in its life cycle.

Even though this data is less important, you'll want to keep it so you can occasionally perform historical analysis on it. Figure 4-18 shows a common data usage trend in analysis situations.

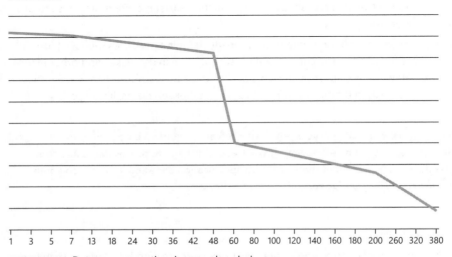

FIGURE 4-18 Data usage over time in several analysis processes

To facilitate how accessible your data is, and how much it costs to store it, you can define an access tier for it. Data access tier options are as follows:

- **Hot** Data will be accessed frequently.
- **Cool** Data will be accessed from time to time.
- **Archive** Data will be kept in the storage, but it is not intended to be accessed.

NOTE Not all your data will be accessed in the same way — and not all your data will be accessed in the same way over time.

Use life cycle management rules

In General Purpose V2 Azure storage accounts, you can use various data-management policies to change how your data is stored. The policies enable you to do the following:

- Automatically move data from the hot tier to the cool tier or from the cool tier to the archive tier if it is not accessed during a specified period.
- Automatically move data from the cool tier to the hot tier when it is accessed.
- Execute movements on a scheduled basis.
- Delete old data.

These policies can be applied to the entire account or to specific containers. Moreover, they can be assigned to blob subsets, defined by part of the name, like its prefix, or by blob index tags. (Blob index tags are assigned to blob content in its metadata.)

PRACTICE   Define a life cycle management rule

In this practice exercise, you will define rules to move data in your data lake from the hot tier to the cool tier. Follow these steps:

1. Open *https://portal.azure.com* and use your credentials to log in.
2. Navigate to your storage account.
3. Click the **Lifecycle Management** option under **Data Lake Storage** in the left panel.

   > **NOTE**   If you want to set life cycle rules for a storage account with no hierarchical namespace enabled, you will find the same option under Blob Storage.

4. Click the **Add a Rule** button at the top of the list view.
5. Type a name for the rule — in this example, CoolDemography.
6. Select the **Limit Blobs with Filters** option.
7. Leave the **Blob Type** and **Blob Subtype** settings as is.
8. Click the **Next** button.
9. In the **Base Blobs** tab, define the **If** condition by entering a value in the **More Than (Days Ago)** box — in this case, **2**.
10. Open the **Then** drop-down list and select **Move to Cool Storage**.
11. Click the **Next** button.
12. In the **Filter Set** tab, enter a prefix match for the filter — in this example, **globaldata/ Demography**.

    This will filter the data for countries' demographics but not the daily information stored in the Daily folder.
13. Click the **Add** button.

Define a life cycle management rule using JSON

Like almost any other element in Azure, rules can also be defined using JSON. To do so, repeat steps 1 through 3 from the previous practice exercise. Then click the **Code View** tab. You'll see JSON code like this:

```json
{
  "rules": [
    {
      "enabled": true,
      "name": "CoolDemography",
      "type": "Lifecycle",
      "definition": {
        "actions": {
          "baseBlob": {
            "tierToCool": {
              "daysAfterModificationGreaterThan": 2
            }
          }
        },
        "filters": {
          "blobTypes": [
            "blockBlob"
          ],
          "prefixMatch": [
            "globaldata/Demography"
          ]
        }
      }
    }
  ]
}
```

You can edit the JSON code to refine the rules. For example, you can change the "prefixMatch" filter to "blobIndexMatch" and define it as a metadata tags filter to match with the name of the tag, the operator to apply, and the value, as in the following code:

```json
{"name": "SourceDate","op": "<","value": "20200930"}
```

You can define the following actions, or rules, in JSON:

- **tierToCool** Change to the cool tier.
- **enableAutoTierToHotFromCool** Move the blob to the hot tier when accessed.
- **tierToArchive** Change the tier to archive.
- **Delete** Remove the blob.

The conditions to execute these rules could be:

- **daysAfterModificationGreaterThan** Last modified day
- **daysAfterCreationGreaterThan** Creation date, regardless of if and when it has been modified
- **daysAfterLastAccessTimeGreaterThan** Last date accessed for reading or writing

In all cases, the value is an integer.

Use PowerShell or CLI for life cycle management

When you need to manage several rules in different storage accounts, you might prefer to automate this procedure. The following script, written in PowerShell, accomplishes this. Notice that it uses the Az library, so it will be easy to migrate this to an Az CLI or create a C# application using the same library.

```
[CmdLetBinding()]
param
(
    [Parameter(Mandatory=$true,Position=0,HelpMessage="You must set the Resource Group" )]
        [string]$ResourceGroupName,
    [Parameter(Mandatory=$true,Position=1,HelpMessage="You must set the Storage Account" )]
        [string]$AccountName,
    [Parameter(Mandatory=$true,Position=2,HelpMessage="The Action is required")]
    [ValidateSet("Delete","TierToArchive","TierToCool")]
        [string]$BlobAction,
    [Parameter(Mandatory=$true,Position=3,HelpMessage="You must Define the days after
modification Value" )]
        [int]$Days,
    [Parameter(Mandatory=$true,Position=4,HelpMessage="You must set the prefix filter" )]
        [string]$Prefix
)


Import-Module -Name Az
#Connect to your Azure Account
Connect-AzAccount


#Create a new action object
$action = Add-AzStorageAccountManagementPolicyAction `
    -BaseBlobAction $BlobAction `
    -daysAfterModificationGreaterThan $days
```

```
# Create a new filter object
$filter = New-AzStorageAccountManagementPolicyFilter `
    -PrefixMatch $Prefix

#Create a new rule object
$rule1 = New-AzStorageAccountManagementPolicyRule `
    -Name Test `
    -Action $action `
    -Filter $filter

#Set the policy
Set-AzStorageAccountManagementPolicy `
    -ResourceGroupName $ResourceGroupName `
    -StorageAccountName $AccountName `
    -Rule $rule1
```

> **NOTE** You can find this script in the companion content in the Lifecycle Management.ps1 file.

Summary

- Azure Monitor is the central service to process information about all your resources and how they are working.
- Azure Monitor uses two sets of information: metrics and logs.
- Metrics are stored as time series, while logs are different for resources and monitoring elements.
- Each resource presents a dashboard with charts of the most critical metrics in its Overview page in Azure Portal.
- You can define your own charts, combining metrics if you wish, and pin them to the existing dashboard or to a new custom-created one.
- You can define the auto-refresh period, time zone, and filters that will apply to all charts in the dashboard.
- You can combine different charts from different resources in Jupyter Notebooks to maintain all the information you need in one place.
- Information usually queried from DMVs in a traditional Azure SQL Database is exposed as metrics for Azure Monitor.
- The metrics for storage accounts are about capacity, size of content, space used, transactions, and others related to the effectiveness of the service, like end-to-end latency and server-processing latency.
- The sole difference between metrics for Azure Blob storage and for ADLS is the existence of the IndexCapacity metric, which makes sense only with a hierarchical namespace for ADLS.

- Metrics for Cosmos DB include those to measure throughput for cost evaluation, consistency, and replication.

- To use alerts in Azure, you must enable Azure Insights Provider.

- Alerts are linked to action groups, making it easier to organize them and to configure the kind of communication to establish.

- Action groups can be assigned to actions, to be performed when they receive alerts.

- Logs are not generated by default. Diagnostic settings must be defined for each resource to generate the desired logs.

- There are two major groups of logs: those related to the service platform, which can be configured on the Activity Log page, and those related to the activities of the resource, managed in the resource's Diagnostic Settings page in Azure Portal.

- To centralize logs in one place for an entire resource group, you can create an Azure Log Analytics workspace.

- Logs can be stored in a storage account or sent to an Event Hub to be processed by external applications.

- To query logs in Azure Log Analytics, you use Kusto Query Language (KQL), which can filter, order, and select information, and express it in different formats.

- Azure Databricks requires the implementation of custom code to send information to Azure Log Analytics, but a prebuilt library and sample codes are available at GitHub.

- Metrics and logs can be queried and processed externally using PowerShell, Azure CLI, the REST API, and .NET libraries. Some external open-source applications, like Grafana, can be used as well.

- Azure Data Factory metrics include counters for pipelines, activities, and triggers. These metrics have different counters for failed runs, successful runs, and canceled runs. A separate set of metrics is available for SSIS execution.

- In the Monitor section of the Data Factory Overview page, you can display pipeline metrics in a Gantt graph, suspend pipelines, or cancel them.

- Azure Databricks does not display metrics per se in Azure Portal. However, there is a feature in preview to add them.

- You can send log information from Azure Databricks to Azure Log Analytics by using a job defined in your implementation with Java or Scala.

- For Azure Stream Analytics, the SUs metric enables you to evaluate how well your pool is working.

- As with SQL pools, you can use metrics to monitor Apache Spark pools and analyze logs to evaluate the tasks processed by them.

- Analyzing Apache Spark logs, you can trace different related operations by their operation IDs.

- You can obtain information from metrics using PowerShell, CLI, the REST API, and .NET libraries. This enables external tools like Grafana to build charts and dashboards outside Azure Portal.

- Information from different resources, metrics, and logs can be consolidated in Azure Log Analytics.

- You can query and make charts from metrics from different resources at the same time in the Azure Log Analytics page in Azure Portal.

- A specialized Jupyter Notebook allows for the centralization of data in complex relationships to immediately view information about the entire data environment.

- For high-volume data operations, there are several ways to partition data: horizontally, vertically, and functionally.

- Some resources, like Cosmos DB, manage partition logic internally. Still, you can have an effect on the service's performance by defining good partitioning logic for your data.

- Optimizing your data environment requires the implementation of a combination of best practices and involves constantly monitoring the entire process. Among these best practices are managing partitions for data ingestion and monitoring Azure Synapse Analytics by using DMVs.

- To avoid excessive bills, it's important to use metrics to identify and evaluate badly assigned resources. Also, having rules relating to data lake storage tiers will reduce the costs of data storage.

Summary exercise

In this summary exercise, test your knowledge about topics covered in this chapter. You can find answers to the questions in this exercise in the next section.

Suppose you work in a consulting company that helps customers optimize their Azure platforms. It's your job to field queries about how to manage each customer's environment. Here are a few examples:

1. A customer decides to move to online sales and wants to implement distributed data across different regions to optimize responses and management. The biggest data movements will involve the orders-management and sales processes. There are millions of daily sales in countries with varying populations and GDP per capita. In addition, product information is very detailed and descriptive, with several paragraphs explaining each product's features. Customer satisfaction is an important metric to calculate the ranking of each offering and to identify trends to enable the customer to display best-ranked products first. Which of the following would be the best implementation for this situation?

   A. Cosmos DB with geo-replication for product storage and a centralized Azure SQL Database to store operations and customer-satisfaction data

   B. Cosmos DB with geo-replication for product storage and an Azure SQL elastic pool, sharded by country, to store all operations and customer-satisfaction data

   C. Cosmos DB with geo-replication for product storage and an Azure SQL elastic pool, sharded by product category, to store all operations and customer-satisfaction data

D. Cosmos DB with geo-replication for product storage; an Azure SQL elastic pool, sharded by product category, to store all operations; and an Azure ADLS Gen2 for cumulative customer-satisfaction data

E. A widely geo-distributed Cosmos DB containing all the information

2. Customer-satisfaction data is analyzed frequently to recalculate rankings and, at the same time, perform customer services in cases when complaints arise. The analysis is implemented in Azure Synapse Analytics. You must evaluate how it is working and define recommendations for further follow-up by the customer team. What is the best way to give information to the customer team?

A. Set up monthly meetings to review the results

B. Train the customer team to query data from the metrics of each resource in use

C. Configure proper diagnostic settings in each resource and send them to Azure Log Analytics

D. Implement option C and prepare a workbook with the most important data from the log analysis, including explanations about how to interact with it and steps to follow to optimize resources when the results are not the best

E. Implement option D and set alerts for the riskiest situations

Summary exercise answers

This section contains the answers to the summary exercise questions in the previous section and explains why each answer is correct.

1. **D: Cosmos DB with geo-replication for product storage; an Azure SQL elastic pool, sharded by product category, to store all operations; and an Azure ADLS Gen2 for cumulative customer-satisfaction data:** Sharding by country will result in irregular distributed data because some countries will generate much more data movement than others. Customer-satisfaction information must be stored in ADLS to perform different analysis over time and could be handled by some kind of life-cycle management if needed.

2. **E: Implement option D and set alerts for the riskiest situations:** Having information in Azure Log Analytics surfaces key information for the team, preserves other data to be analyzed in the event of a problem, and enhances the evaluation process. Moreover, if a problem occurs, you can easily add a query or metric and related explanations to a workbook. Defining alerts prevents a situation in which the customer team forgets to review the workbook in a timely manner. If something is wrong, the team will be advised.

Index

A

ABFS drivers, 44
Access Control Lists (ACL), ADLS Gen2, 44
access tiers, Blob storage, 9
accessing data, security, 74–75
    Azure AD, 75–78
    Cosmos DB, 80
    GRS, 83–84
    GZRS, 84
    RBAC, 78–80
    SAS, 81–82
    ZRS, 83
accounts
    Azure storage accounts, creating to use data lakes, 44–45
    Cosmos DB, 35
        creating accounts, 38–39
        diagnostic settings, 238–239
acquisition, data streams, 189–190
action groups, defining, 233
Active Directory (AD)
    Azure SQL database, 93–94
    data access security, 75–78
activities, Azure Data Factory, 125
advanced analytics, use cases, 26
aggregation, data streams, 190, 192
alerts
    Azure Data Factory, 246–247
    Azure Monitor, 232, 255–256
    metrics, 233–236
analysis engines, ADLS Gen2, 44
analytics
    advanced analytics, use cases, 26
    Azure Databricks, 25
    Azure Log Analytics, 236
        audit logs, 239–242
        audits, 256–260
        log configuration, 238
        notebooks, 257–260
        server-level auditing, 237
        workspace preparation, 236–237
    Azure SQL Analytics, deploying, 242–244
    Azure SQL Managed Instance (SQL MI), 14
    Azure Stream Analytics, 22–23, 46
        event processing, 192–193
        jobs, 193–212
        logs, 250
        metrics, 249
        monitoring, 249–250
        optimizing, 268
    Azure Synapse Analytics, 15–16, 46
        components of, 15
        data distribution, 100–101
        Dedicated SQL Pools, 15, 74
        monitoring, 229–230, 250–255
        optimizing, 269–270
        Parquet data types, 70–71
        provisioning workspaces, 70–74
        replication strategies, 270
        scaling, 99–100
        Serverless Spark Pools, 15
        Serverless SQL Pools, 15
        Spark, 15, 73–74
        Synapse pipelines, 15
        Synapse SQL, 15
        Synapse Studio, 15
        table partitions, 101–102
Apache Spark, 15
    Azure Databricks, 25
    data frames, 128
    Graph X, 128
    ingesting data, Azure Databricks, 168–176
    MLlib, 128
    pools, 185–189
        adding to Azure synapse Analytics workspaces, 73–74
        cluster management, 271–272
        data management, 270
        logs, 254–255
        memory management, 271
        monitoring, 253–255
        optimizing, 270–272
    PySpark
        notebooks, 187–189
        reading data, 187
    RDD, 127
    Serverless pools, 15
    Spark Core API, 127
    SQL, 128
    streaming, 128
Apache TinkerPop Gremlin API, 36
apiVersion entry, ARM templates, 53
application design, partitions, 262–263
Application Programming Interface (API)
    Azure Table API, 36
    Cassandra API, 35–36
    Core (SQL) API, 35
    Cosmos DB, 35–36
    Gremlin API, 36
    MongoDB API, 35
    REST API, ADLS Gen2, 12, 44
    uploading data to storage, 66–67
architectural diagrams, Synapse SQL pools, 15–16
archive access tier, Blob storage, 9
ARM templates, 50
    analyzing, 51–54
    apiVersion entry, 53
    automated resource deployments, 57
        .NET framework, 59–60
        command line, 59

PowerShell, 57–58
dependsOn entry, 53
downloading from Azure Portal, 51
entries, 53–54
kind entry, 53
location entry, 53
name entry, 53
properties entry, 54
sku entry, 53
storage, 54–56
type entry, 53
audits
Azure Log Analytics, 236, 256–260
audit logs, 239–242
log configuration, 238
server-level auditing, 237
workspace preparation, 236–237
server-level auditing, Azure Log Analytics, 237
authentication, AD, Azure Data Studio, 93–94
authoring
Azure Data Factory, 122–124
jobs, 194–195
notebooks, 164
authorization
data access security
Azure AD, 75–78
Cosmos DB, 80
RBAC, 78–80
SAS, 81–82
automated resource deployments
ARM templates, 57
.NET framework, 59–60
command line, 59
PowerShell, 57–58
autoscaling, 166–167
availability, partitions, 262
AzCopy, uploading data to storage, 65–66
Azure AD (Active Directory), data access security, 75–78
Azure Blob storage, 9, 10–11
access tiers, 9
ADLS Gen1, 11
ADLS Gen2, 11
cost, 12
data redundancy, 12
features of, 11–12
hadoop-compatible access, 11

hierarchical namespaces, 11
ingestion tools, 12
optimized performance, 12
reference inputs, 197–198
REST API, 12
security, 11
stream inputs, 197
using, 12
archive access tier, 9
Azure Cosmos DB, 16–18
Azure SQL database, 13
deployment options, 13
DTU model, 13
elastic pool deployments, 13
managed instance deployments, 13
models of, 13–14
single database deployments, 13
using, 14
vCore model, 14
Azure Synapse Analytics, 15–16
components of, 15
Dedicated SQL Pools, 15
Serverless Spark Pools, 15
Serverless SQL Pools, 15
Spark, 15
Synapse pipelines, 15
Synapse SQL, 15
Synapse Studio, 15
containers, 9
cool access tier, 9
GRS, 10
GZRS, 10
hot access tier, 9
LRS, 9
monitoring, 226
logs, 228–229
metrics, 226–228
partitioning, 265
redundancy, 9–10
Snapshots for Blobs, 85
SQL MI, 14
ZRS, 10
Azure Cosmos DB. See Cosmos DB
Azure Data Box, 45
Azure Data Explorer, 45
Azure Data Factory (ADF), 23–24, 45–46
activities, 125
alerts, 246–247
authoring, 122–124

batch processing, 119–125, 135–136
components of, 124–125
connectors
categories, 148
creating, 147–152
updating, 148
Copy Data wizard, creating pipelines, 152–157
creating, 120–122
datasets, 125
creating, 147–152
DIU, 136
hybrid ETL with existing in-premises SSIS, 26–27
IR, 136–137
configuring, 137
managing, 137
linked services, 125
logs, 246
metrics, 244–246
parameters, 143
pipelines, 125
attaching triggers to, 161–162
creating, 142–147, 152–157
monitoring, 244–247
triggers
attaching to pipelines, 161–162
creating, 157–159
event triggers, 161
scheduling, 159–160
tumbling window triggers, 160–161
Azure Data Lake Storage (ADLS)
ADLS Gen1, 11, 44
ADLS Gen2, 11, 43–44
ABFS drivers, 44
ACL permissions, 44
analysis engines, 44
Azure Portal, 44
Azure services with access to ADLS, 45–46
Azure storage accounts, creating to use data lakes, 44–45
cost, 12
data redundancy, 12
features of, 11–12
hadoop-compatible access, 11
HDFS, 44
hierarchical namespaces, 11
ingestion tools, 12

optimizing, 12, 267
POSIX permissions, 44
REST API, 12, 44
security, 11
using, 12
monitoring, 229
Azure Data Studio, Azure SQL
database connections, 89–96
Azure Databricks, 24–25, 46
autoscaling, 166–167
batch processing, 119, 125–136
billing, 133–135
characteristics of, 126–128
Cluster Managers, 126
clusters
creating, 162–163, 166–167
provisioning, 164
control plane, 130–131
data frames, 128
DBFS, 132–133
DBU, 133–135
Delta Lakes, 126
enterprise-level security, 126
Graph X, 128
ingesting data, 168–176
instances
navigating, 130–135
provisioning, 128–130
jobs, creating, 164–165
MLlib, 128
MLOps, 126
/mnt folders, 133
monitoring, 247–249
notebooks, 126
authoring, 164
exporting, 167
scheduling, 164
optimized runtime, 126
RDD, 127
REST API, 127
Spark Core API, 127
Spark SQL, 128
streaming, 128
workspaces, creating, 165–168
Azure Event Grid, 46
Azure Event Hub, 46
Azure HDInsight, 46
Azure Insights Provider
action groups, defining, 233
enabling, 232
Azure IoT Hub, 46
Azure IR (Integrated Runtimes),
138–139
Azure Log Analytics, 236

audits, 236, 256–260
logs, 238, 239–242
server-level auditing, 237
workspace preparation,
236–237
logs
audit logs, 239–242
configuring, 238
notebooks, 257–260
server-level auditing, 237
workspace preparation, 236–237
Azure Logic Apps, 46
Azure Machine Learning, 46
Azure Monitor
alerts, 232, 255–256
data storage, 220
logs, 219
metrics, 219
Azure Portal
ADLS Gen1, 44
ARM templates, downloading, 51
consistency levels, changing, 49
Azure SQL Analytics, deploying,
242–244
Azure SQL Data Warehouse. *See*
Azure Synapse analytics
Azure SQL database, 13
AD authentication, 93–94
Azure Data Studio connections,
89–96
connection security, 96
creating, 88–89
databases, 92–93
deployment options, 13
disaster recovery, 97–99
DTU model, 13
elastic pools, 13, 263–265
geo-replication, 99
HA, 97–99
logs, 224–225
managed instance deployments,
13
metrics, 222–224
models of, 13–14
monitoring, 220–225
partitioning, 263–265
passwords, 92–94
PolyBase, 102–104
shardlets, 263–265
single database deployments, 13
SLA, 97
user names, 92–94
using, 14
vCore model, 14

Azure SQL Managed Instance (SQL
MI), 14
Azure-SSIS IR, 136, 141–142
Azure storage accounts, creating to
use data lakes, 44–45
Azure Stream Analytics, 22–23,
25, 46
event processing, 192–193
jobs, 193
authoring, 194–195
components of, 193
configuring inputs/outputs,
196–201
creating, 207–212
editing queries, 206
provisioning, 193–195
selecting built-in functions,
201–203
monitoring, 249
logs, 250
metrics, 249
optimizing, 268
windows, 202–203
hopping windows, 203–204
session windows, 205–206
sliding windows, 204–205
tumbling windows, 203
Azure Synapse Analytics, 15–16, 46
components of, 15
data distribution, 100–101
Dedicated SQL Pools, 15, 74
ingesting data, 176–178
Spark pools, 185–189
SQL pools, 178–185
metrics, 250
monitoring, 229–230, 250
optimizing, 269–270
Parquet data types, 70–71
processing data, 176–178
Spark pools, 185–189
SQL pools, 178–185
provisioning workspaces,
70–74
replication strategies, 270
scaling, 99–100
Serverless Spark Pools, 15
Serverless SQL Pools, 15
Spark, 15, 73–74
Synapse pipelines, 15
Synapse SQL, 15
Synapse Studio, 15
table partitions, 101–102
Azure Table API, 36

B

batch processing, 19, 115–116, 118–119
  Azure Data Factory, 119–125, 135–136
  Azure Databricks, 119, 125–136
  ELT data extraction, 116–117
  ETL data extraction, 116–117
  Kappa architectures, 21
  Lambda architectures, 20–21
billing, Azure Databricks, 133–135
Blob storage, 9, 10–11
  access tiers, 9
  ADLS Gen1, 11
  ADLS Gen2, 11
    cost, 12
    data redundancy, 12
    features of, 11–12
    hadoop-compatible access, 11
    hierarchical namespaces, 11
    ingestion tools, 12
    optimized performance, 12
    reference inputs, 197–198
    REST API, 12
    security, 11
    stream inputs, 197
    using, 12
  archive access tier, 9
  Azure Cosmos DB, 16–18
  Azure SQL database, 13
    deployment options, 13
    DTU model, 13
    elastic pool deployments, 13
    managed instance deployments, 13
    models of, 13–14
    single database deployments, 13
    using, 14
    vCore model, 14
  Azure Synapse Analytics, 15–16
    components of, 15
    Dedicated SQL Pools, 15
    Serverless Spark Pools, 15
    Serverless SQL Pools, 15
    Spark, 15
    Synapse pipelines, 15
    Synapse SQL, 15
    Synapse Studio, 15
  containers, 9
  cool access tier, 9
  GRS, 10
  GZRS, 10
  hot access tier, 9
  LRS, 9
  monitoring, 226
    logs, 228–229
    metrics, 226–228
  partitioning, 265
  redundancy, 9–10
  Snapshots for Blobs, 85
  SQL MI, 14
  ZRS, 10
bounded staleness consistency level, 48
built-in functions, selecting for jobs, 201–203

C

Cassandra API, 35–36
changing
  ARM templates, storage, 54–56
  consistency levels, 49
CLI, data life cycle management, 275–276
Cluster Managers, 126
clusters
  creating, 162–163, 166–167
  managing, Spark pools, 271–272
  provisioning, 164
cold data-processing paths, Lambda architectures, 20
collections. See containers
column families, 6–7
columnar data stores, 6–7
command line, ARM template resource deployments, 59
common serving layer, Lambda architectures, 20
configuring
  alerts, Azure Monitor, 255–256
  Azure Log Analytic logs, 238
  consistency levels in Cosmos DB, 50
  geospatial configuration, containers, 41
  IR, 137
connectors
  categories, 148
  creating, 147–152
  updating, 148
consistency levels
  bounded staleness, 48
  changing, 49
  configuring in Cosmos DB, 50
  consistent prefix, 48
  Cosmos DB, 48–50
  eventual, 48
  session, 48, 49–50
  strong, 48
consistent prefix consistency level, 48
consumer groups, stream-transport and processing engines, 191–192
containers
  Blob storage, 9
  Cosmos DB, 36
    creating containers, 39–40
    working with information in Data Explorer, 40–41
  functions, 41
  geospatial configuration, 41
  indexing policies, 41
  partition keys, 41
  stored procedures, 41
  triggers, 41
  TTL, 41
control nodes, Synapse SQL pools, 15
cool access tier, Blob storage, 9
Copy Data wizard, pipeline creation, 152–157
Core (SQL) API, 35
Cosmos DB, 16–18, 34–35
  accounts, 35
    creating, 38–39
    diagnostic settings, 238–239
  API, 35–36
    Azure Table API, 36
    Cassandra API, 35–36
    Core (SQL) API, 35
    Gremlin API, 36
    MongoDB API, 35
  consistency levels, 48–50
  containers, 36
    creating, 39–40
    working with information in Data Explorer, 40–41
  data access security, 80
  databases, 36
  global distributions, 85–86
  HA, 86–87
  monitoring, 231–232
  partitioning, 266
  partitons, 46–48
  programming, 41–42
    obtaining RU information, 42–43
    running queries, 42–43

redundancy, 85–86
resource model, 36
RU, 37, 42–43
storage management, 36
uploading data to storage,
62–63
cost, ADLS Gen2, 12
creating, Azure Data Factory,
120–122

D

data access security, 74–75
Azure AD, 75–78
Cosmos DB, 80
GRS, 83–84
GZRS, 84
RBAC, 78–80
SAS, 81–82
ZRS, 83
data at rest, encryption, 107–109
data distribution
Azure Synapse Analytics,
100–101
hash distributions, 100
replicated table distributions, 101
round robin distributions, 101
data encryption
data at rest, 107–109
data in motion, 109
data engineers
implementation tasks, 1
responsibilities, 1
data extraction
ELT, 1
ETL, 1, 26–27
data frames, Spark, 128
data in motion, encryption, 109
data ingestion
Azure Databricks, 168–176
Azure Synapse Analytics, 176–178
Spark pools, 185–189
SQL pools, 178–185
data lakes
ADLS Gen1, 11, 44
ADLS Gen2, 11, 43–44
ABFS drivers, 44
ACL permissions, 44
analysis engines, 44
Azure Portal, 44
Azure services with access to
ADLS, 45–46

Azure storage accounts,
creating to use data lakes,
44–45
cost, 12
data redundancy, 12
features of, 11–12
hadoop-compatible access, 11
HDFS, 44
hierarchical namespaces, 11
ingestion tools, 12
optimizing, 12, 267
POSIX permissions, 44
REST API, 12, 44
security, 11
using, 12
monitoring, 229
data management
life cycles, 272
CLI, 275–276
PowerShell, 275–276
rules, 273–275
security, 104–107
Spark pools, 270
data masking, dynamic, 104–107
data optimization, 260
ADLS Gen2, 267
partitions, 260–263
data partitions
application design, 262–263
availability, 262
Azure SQL database, 263–265
Blob storage, 265
Cosmos DB, 266
functional partitioning, 261
horizontal partitioning
(sharding), 260
query performance, 261–262
scalability, 261
troubleshooting, 260–263
vertical partitioning, 261
data processing, 18
ADF, 23–24
Azure Databricks, 24–25
Azure Stream Analytics, 22–23,
25
Azure Synapse Analytics, 176–178
Spark pools, 185–189
SQL pools, 178–185
batch processing, 19, 115–116,
118–119
Azure Data Factory, 119–125,
135–136
Azure Databricks, 119,
125–136

ELT data extraction, 116–117
ETL data extraction, 116–117
Kappa architectures, 21
Lambda architectures, 20–21
monitoring, 244
stream processing, 19–20,
115–116
Kappa architectures, 21
Lambda architectures, 20–21
Data Query Editor, dynamic data
masking, 106
data redundancy, ADLS Gen2, 12
data storage
account redundancy, 82–84
ADLS Gen2, 43–44
ABFS drivers, 44
ACL permissions, 44
analysis engines, 44
Azure Portal, 44
Azure services with access to
ADLS, 45–46
Azure storage accounts,
creating to use data lakes,
44–45
HDFS, 44
POSIX permissions, 44
REST API, 44
API, uploading data to storage,
66–67
AzCopy, uploading data to
storage, 65–66
Blob storage, 9, 10–11
access tiers, 9
ADLS Gen1, 11
ADLS Gen2, 11–12
archive access tier, 9
Azure Cosmos DB, 16–18
Azure SQL database, 13–14
Azure Synapse Analytics, 15
containers, 9
cool access tier, 9
GRS, 10
GZRS, 10
hot access tier, 9
LRS, 9
redundancy, 9–10
Snapshots for Blobs, 85
SQL MI, 14
ZRS, 10
columnar data stores, 6–7
Cosmos DB, 34–35
account diagnostic settings,
238–239
accounts, 35, 38–39

API, 35–36
Azure Table API, 36
Cassandra API, 35–36
consistency levels, 48–50
containers, 36
containers, creating, 39–40
containers, working with information in Data Explorer, 40–41
Core (SQL) API, 35
databases, 36
Gremlin API, 36
MongoDB API, 35
monitoring, 231–232
obtaining RU information, 42–43
partitioning, 46–48, 266
programming, 41–43
resource model, 36
RU, 37
running queries, 42–43
storage management, 36
uploading data to storage, 62–63
disaster recovery, 84–85
document data stores, 5–6
graph data stores, 7–8
HA, 82–84
key/value data stores, 6
LRS, 83
Microsoft.Azure.Storage.Blob, uploading data to storage, 68
monitoring, 220
non-relational (NoSQL)
databases, 5
columnar data stores, 6–7
document data stores, 5–6
graph data stores, 7–8
key/value data stores, 6
non-relational data storage. See Cosmos DB
relational data stores, 88
Azure SQL database, 88–96
data access security, 88–104
relational databases, 4–5
semi-structured data, 2–3
storage accounts, 8–9
Storage Explorer, uploading data to storage, 63–65
structured (relational) data, 2
unstructured data, 3
data streams
acquisition, 189–190
aggregation, 190, 192

pipelines, 189
process overview, 190
production, 189
storage, 190
stream-transport and processing engines, 191–192
transformation, 190
data, types of, 2
semi-structured data, 2–3
structured (relational) data, 2
unstructured data, 3
databases
Azure SQL database, 92–93
AD authentication, 93–94
Azure Data Studio connections, 89–96
connection security, 96
creating, 88–89
disaster recovery, 97–99
geo-replication, 99
HA, 97–99
passwords, 92–94
PolyBase, 102–104
SLA, 97
user names, 92–94
Cosmos DB, 36
non-relational (NoSQL)
databases, 5
columnar data stores, 6–7
document data stores, 5–6
graph data stores, 7–8
key/value data stores, 6
RDBMS, 4–5
relational databases, 4–5
datasets
Azure Data Factory, 125
creating, 147–152
RDD, 127, 270
DBFS (Databricks File System), 132–133
DBU (Databricks Units), 133–135
Dedicated SQL Pools, 15, 74
deletes, soft, 84
Delta Lakes, 25, 126
dependsOn entry, ARM templates, 53
deploying
ARM templates, 57
.NET framework, 59–60
command line, 59
PowerShell, 57–58
Azure SQL Analytics, 242–244
Azure SQL database, 13

diagnostics, Cosmos DB account settings, 238–239
disaster recovery
Azure SQL database, 97–99
data storage, 84–85
soft deletes, 84
distributions (global), Cosmos DB, 85–86
DIU (Data Integration Units), 136
DMS (Data Movement Service), Synapse SQL pools, 15
DMV (Dynamic Management Views)
monitoring, SQL pools, 250–253
monitoring Azure SQL Database, 225
document data stores, 5–6
documents, JSON, 5–6
downloading ARM templates, 51
DTU model, Azure SQL database, 13
dynamic data masking, 104–107

E

edges, defined, 7
editing queries (jobs), 206
elastic pools, 13, 263–265
ELT data extraction, 1, 116–117
encryption, data
data at rest, 107–109
data in motion, 109
enterprise-level security, Azure Databricks, 126
ETL data extraction, 1
batch processing, 116–117
hybrid ETL with existing in-premises SSIS, 26–27
Event Hub stream inputs, 196–197
event processing, 192–193
event triggers, 161
eventual consistency level, 48
exporting notebooks, 167
external tables, SQL pools, 178–181
extracting data
ELT, 1
ETL, 1, 26–27

F

file systems, HDFS, 44
functional partitioning, 261
functions, containers, 41

G

geo-replication, Azure SQL
database, 99
geospatial configuration, containers,
41
global distributions, Cosmos DB,
85–86
graph data stores, 7–8
Graph X, Spark, 128
Gremlin API, 36
GRS (Geo-Redundant Storage), 10,
83–84
GZRS (Geo-Zone-Redundant
Storage), 10, 84

H

HA (High Availability)
Azure SQL database, 97–99
Cosmos DB, 86–87
data storage, 82–84
hadoop-compatible access, ADLS
Gen2, 11
hash distributions, 100
HDFS (Hadoop Distributed File
Systems), 44
hierarchical namespaces, ADLS
Gen2, 11
hopping windows, 203–204
horizontal partitioning (sharding),
260–261
hot access tier, Blob storage, 9
hot data-processing paths, Lambda
architectures, 20
hybrid ETL with existing in-premises
SSIS, 26–27

I

indexing policies, containers, 41
ingesting data
Azure Databricks, 168–176
Azure Synapse Analytics, 176–178
Spark pools, 185–189
SQL pools, 178–185
ingestion tools, ADLS Gen2, 12
inputs (jobs)
Blob Storage/ADLS Gen 2
reference inputs, 197–198

Blob Storage/ADLS Gen 2 stream
inputs, 197
configuring, 196–199
Event Hub stream inputs,
196–197
IoT Hub stream inputs, 197
SQL Database reference inputs,
199
Intelligent Insights, 226
IoT (Internet of Things),
architectures, 28
IoT Hub stream inputs, 197
IR (Integrated Runtimes), 136
Azure IR, 136, 138–139
Azure-SSIS IR, 136, 141–142
configuring, 137
creating, 138–142
DIU, 136
managing, 137
self-hosted IR, 136, 139–140

J

jobs
authoring, 194–195
Azure Stream Analytics, 193–212
Blob Storage/ADLS Gen 2
reference inputs, 197–198
Blob Storage/ADLS Gen 2 stream
inputs, 197
built-in functions, selecting,
201–203
components of, 193
configuring
inputs, 196–199
outputs, 199–201
creating, 164–165, 207–212
Event Hub stream inputs,
196–197
IoT Hub stream inputs, 197
provisioning, 193–195
queries, editing, 206
SQL Database reference inputs,
199
JSON (JavaScript Object Notation)
Azure Databricks, ingesting data,
168–176
data life cycle management,
274–275
documents, 5–6
semi-structured data, 2–3

K

Kappa architectures, data
processing, 21
key/value data stores, 6
kind entry, ARM templates, 53
Kryo, 271

L

Lambda architectures, data
processing, 20–21
life cycles, data management, 272
CLI, 275–276
JSON and rule definitions,
274–275
PowerShell, 275–276
rules, 273–275
linked services, Azure Data Factory,
125
location entry, ARM templates, 53
logical architectural diagrams,
Synapse SQL pools, 15–16
logs, 219
Azure Data Factory, 246
Azure Log Analytics, 236
audit logs, 236–237
log configuration, 238
server-level auditing, 237
workspace preparation,
236–237
Azure SQL Database, 224–225
Azure Stream Analytics, 250
Blob storage, 228–229
Spark pools, 254–255
LRS (Locally Redundant Storage),
9, 83

M

managed instance deployments,
Azure SQL database, 13
managing
clusters, Spark pools, 271–272
data
life cycles, 272–276
security, 104–107
Spark pools, 270
IR, 137
memory, Spark pools, 271

masking data (dynamic), 104–107
memory management, Spark pools, 271
message-ingestion engines, 191
metrics, 219
    alerts, 233–236
    Azure Data Factory, 244–246
    Azure SQL Database, 222–224
    Azure Stream Analytics, 249
    Azure Synapse Analytics, 250
    Blob storage, 226–228
Microsoft Power BI, 46
Microsoft.Azure.Storage.Blob, uploading data to storage, 68
MLlib, Spark, 128
MLOps, Azure Databricks, 126
/mnt folders, Azure Databricks, 133
MongoDB API, 35
monitoring
    ADLS, 229
    alerts, 232, 255–256
    Azure Databricks, 247–249
    Azure SQL Database, 220–222
        DMV monitoring, 225
        logs, 224–225
        metrics, 222–224
    Azure Stream Analytics, 249
        logs, 250
        metrics, 249
    Azure Synapse Analytics, 229–230, 250
    Blob storage, 226
        logs, 228–229
        metrics, 226–228
    Cosmos DB, 231–232
    data processing, 244
    data storage, 220
    pipelines, Azure Data Factory, 244–247
    Spark pools, 253–255
    SQL pools with DMV, 250–253
motion, data in, 109
MPP, Synapse SQL pools, 15

N

name entry, ARM templates, 53
namespaces, hierarchical namespaces, ADLS Gen2, 11
navigating Azure Databrick instances, 130–135

.NET framework, ARM templates, resource deployments, 59–60
nodes, defined, 7
non-relational data storage
    ADLS Gen2, 43–44
        ABFS drivers, 44
        ACL permissions, 44
        analysis engines, 44
        Azure Portal, 44
        Azure services with access to ADLS, 45–46
        Azure storage accounts, creating to use data lakes, 44–45
        HDFS, 44
        POSIX permissions, 44
        REST API, 44
    Cosmos DB, 34–35
        account diagnostic settings, 238–239
        accounts, 35
        accounts, creating, 38–39
        API, 35–36
        Azure Table API, 36
        Cassandra API, 35–36
        consistency levels, 48–50
        containers, 36
        containers, creating, 39–40
        containers, working with information in Data Explorer, 40–41
        Core (SQL) API, 35
        data access security, 80
        databases, 36
        Gremlin API, 36
        MongoDB API, 35
        monitoring, 231–232
        obtaining RU information, 42–43
        partitioning, 266
        partitons, 46–48
        programming, 41–43
        resource model, 36
        RU, 37
        running queries, 42–43
        storage management, 36
    provisioning
        uploading data to storage, 61–68
        with ARM templates, 50, 51–60
    types of storage serviced by Azure platform, 33–34
NoSQL (non-relational) databases, 5

columnar data stores, 6–7
document data stores, 5–6
graph data stores, 7–8
key/value data stores, 6
notebooks, 126
    authoring, 164
    Azure Log Analytics, 257–260
    exporting, 167
    PySpark notebooks, 187–189
    scheduling, 164

O

OPENROWSET command, SQL pools, 182–185
optimizing
    ADLS Gen2
        data, 267
        performance, 12
    Azure Databricks, 126
    Azure Stream Analytics, 268
    Azure Synapse Analytics, 269–270
    Spark pools, 270–272
outputs (jobs), configuring, 199–201

P

parameters, Azure Data Factory, 143
Parquet data types, Azure Synapse Analytics, 70–71
partitions
    application design, 262–263
    availability, 262
    Azure SQL database, 263–265
    Blob storage, 265
    Cosmos DB, 46–48, 266
    functional partitioning, 261
    horizontal partitioning (sharding), 260
    keys, 41
    query performance, 261–262
    scalability, 261
    table partitions, Azure Synapse Analytics, 101–102
    troubleshooting, 260–263
    vertical partitioning, 261
passwords, Azure SQL database, 92–94
performance, ADLS Gen2, 12
permissions (ACL), ADLS Gen2, 44

pipelines
    Azure Data Factory, 125, 244–247
    creating, 142–147, 152–157
    streaming data, 189
    Synapse, 15
    triggers, attaching to pipelines, 161–162
PolyBase, 102–104
pools
    elastic pools, 263–265
    Spark pools, 185–189
        adding to Azure synapse Analytics workspaces, 73–74
        cluster management, 271–272
        data management, 270
        logs, 254–255
        memory management, 271
        monitoring, 253–255
        optimizing, 270–272
    SQL pools, 178–185
        external tables, 178–181
        monitoring with DMV, 250–253
        OPENROWSET command, 182–185
Portable Operating System Interface (POSIX), permissions, 44
PowerShell
    ARM templates, resource deployments, 57–58
    data life cycle management, 275–276
processing data, 18
    ADF, 23–24
    Azure Databricks, 24–25
    Azure Stream Analytics, 22–23, 25
    Azure Synapse Analytics, 176–178
        Spark pools, 185–189
        SQL pools, 178–185
    batch processing, 19, 115–116, 118–119
        Azure Data Factory, 119–125, 135–136
        Azure Databricks, 119, 125–136
        ELT data extraction, 116–117
        ETL data extraction, 116–117
        Kappa architectures, 21
        Lambda architectures, 20–21
    monitoring, 244
    stream processing, 19–20, 115–116

    Kappa architectures, 21
    Lambda architectures, 20–21
processing events, 192–193
production, data streams, 189
programming Cosmos DB, 41–42
    obtaining RU information, 42–43
    running queries, 42–43
properties entry, ARM templates, 54
provisioning
    Azure Databrick instances, 128–130
    clusters, 164
    jobs, 193–195
    non-relational data stores
        uploading data to storage, 61–68
        with ARM templates, 50, 51–60
PySpark
    notebooks, 187–189
    reading data, 187

Q

queries
    Cosmos DB, running queries, 42–43
    editing, 206
    metrics, 43
    partitions for query performance, 261–262

R

recovery, disaster
    Azure SQL database, 97–99
    data storage, 84–85
    soft deletes, 84
Recovery Point Objectives (RPO), 84
redundancy
    ADLS Gen2, 12
    Blob storage, 9–10
    Cosmos DB, 85–86
    data storage, 82–84
    LRS, 83
relational data stores, 88–96
Relational Database Management Systems (RDBMS), 4–5
relational databases, 4–5
relational (structured) data, 2
replication

Azure Synapse Analytics, 270
    table distributions, 101
Request Units (RU), Cosmos DB, 37, 42–43
Resilient Distributed Datasets (RDD), 127, 270
resource deployments, ARM templates, 57
    command line, 59
    .NET framework, 59–60
    PowerShell, 57–58
resource models, Cosmos DB, 36
rest, data at, 107–109
REST API
    ADLS Gen2, 12, 44
    Azure Databricks, 127
retention periods, 191
Role-Based Access Control (RBAC), data access security, 78–80
round robin distributions, 101
rules, data life cycle management, 273–275
runtime (optimized), Azure Databricks, 126

S

scalability, partitions, 261
scaling Azure Synapse Analytics, 99–100
scheduling
    notebooks, 164
    triggers, 159–160
security
    ADLS Gen2, 11
    Azure SQL database, connection security, 96
    connection security, Azure SQL database, 96
    data access, 74–75
        Azure AD, 75–78
        Cosmos DB, 80
        GRS, 83–84
        GZRS, 84
        RBAC, 78–80
        SAS, 81–82
        ZRS, 83
    data encryption
        data at rest, 107–109
        data in motion, 109
    data management, 104–107

self-hosted IR (Integrated Runtimes),
    136, 139–140
semi-structured data, 2–3
Serverless Spark Pools, 15
Serverless SQL Pools, 15
server-level auditing, Azure Log
    Analytics, 237
Service Level Agreements (SLA),
    Azure SQL database, 97
session consistency level, 48, 49–50
session windows, 205–206
sharding (horizontal partitioning),
    260–261
shardlets, 263–265
Shared Access Signatures (SAS),
    81–82
single database deployments, Azure
    SQL database, 13
sku entry, ARM templates, 53
sliding windows, 204–205
Snapshots for Blobs, 85
soft deletes, 84
Spark, 15
    Azure Databricks, 25
    data frames, 128
    Graph X, 128
    ingesting data, Azure Databricks,
        168–176
    MLlib, 128
    pools, 185–189
        adding to Azure synapse
            Analytics workspaces,
            73–74
        cluster management, 271–272
        data management, 270
        logs, 254–255
        memory management, 271
        monitoring, 253–255
        optimizing, 270–272
    PySpark
        notebooks, 187–189
        reading data, 187
    RDD, 127
    Serverless pools, 15
    Spark Core API, 127
    SQL, 128
    streaming, 128
SQL (Structured Query Language)
    Azure SQL Analytics, deploying,
        242–244
    Azure SQL database, 13
        Azure Data Studio
            connections, 89–96
        creating, 88–89

deployment options, 13
DMV monitoring, 225
DTU model, 13
elastic pools, 13, 263–265
logs, 224–225
managed instance
    deployments, 13
metrics, 222–224
models of, 13–14
monitoring, 220–225
partitioning, 263–265
shardlets, 263–265
single database deployments,
    13
using, 14
vCore model, 14
Core (SQL) API, 35
pools, 178–185
    Dedicated SQL Pools, 15, 74
    external tables, 178–181
    OPENROWSET command,
        182–185
    Serverless SQL Pools, 15
    Synapse SQL pools, 15–16
Spark, 128
SQL Database reference inputs,
    199
SQL MI, 14
SSIS, hybrid ETL with existing in-
    premises SSIS, 26–27
Synapse SQL, 15–16
SQL Server Integration Services
    (SSIS), 26–27
Storage Explorer, uploading data to
    storage, 63–65
stored procedures, containers, 41
storing data
    account redundancy, 82–84
    ADLS Gen2, 43–44
        ABFS drivers, 44
        ACL permissions, 44
        analysis engines, 44
        Azure Portal, 44
        Azure services with access to
            ADLS, 45–46
        Azure storage accounts,
            creating to use data lakes,
            44–45
        HDFS, 44
        POSIX permissions, 44
        REST API, 44
    ADLS, monitoring, 229
    API, uploading data to storage,
        66–67

ARM templates, 54–56
AzCopy, uploading data to
    storage, 65–66
Blob storage, 9, 10–11
    access tiers, 9
    ADLS Gen1, 11
    ADLS Gen2, 11–12
    archive access tier, 9
    Azure SQL database, 13–14
    Azure Synapse Analytics, 15
    containers, 9
    cool access tier, 9
    Cosmos DB, 16–18
    GRS, 10
    GZRS, 10
    hot access tier, 9
    logs, 228–229
    LRS, 9
    metrics, 226–228
    monitoring, 226–229
    partitioning, 265
    redundancy, 9–10
    Snapshots for Blobs, 85
    SQL MI, 14
    ZRS, 10
columnar data stores, 6–7
Cosmos DB, 34–35
    account diagnostic settings,
        238–239
    accounts, 35
    accounts, creating, 38–39
    API, 35–36
    Azure Table API, 36
    Cassandra API, 35–36
    consistency levels, 48–50
    containers, 36
    containers, creating, 39–40
    containers, working with
        information in Data
        Explorer, 40–41
    Core (SQL) API, 35
    databases, 36
    Gremlin API, 36
    MongoDB API, 35
    monitoring, 231–232
    obtaining RU information,
        42–43
    partitioning, 266
    partitons, 46–48
    programming, 41–43
    resource model, 36
    RU, 37
    running queries, 42–43
    storage management, 36

uploading data to storage,
    62–63
data streams, 190
disaster recovery, 84–85
document data stores, 5–6
graph data stores, 7–8
HA, 82–84
key/value data stores, 6
LRS, 83
Microsoft.Azure.Storage.Blob,
    uploading data to storage, 68
monitoring, 220
non-relational (NoSQL)
    databases, 5
    columnar data stores, 6–7
    document data stores, 5–6
    graph data stores, 7–8
    key/value data stores, 6
non-relational data storage. See
    Cosmos DB
relational data stores, 88
    Azure SQL database, 88–96
    data access security, 88–104
relational databases, 4–5
semi-structured data, 2–3
storage accounts, 8–9
Storage Explorer, uploading data
    to storage, 63–65
structured (relational) data, 2
unstructured data, 3
uploading data to storage,
    59–60
stream processing, 19–20, 115–116
    Kappa architectures, 21
    Lambda architectures, 20–21
streaming
    data
        acquisition, 189–190
        aggregation, 190, 192
        pipelines, 189
        process overview, 190
        production, 189
        storage, 190
        stream-transport and
            processing engines,
            191–192
        transformation, 190
    Spark, 128

stream-transport and processing
    engines, 191–192
strong consistency level, 48
structured (relational) data, 2
Synapse
    pipelines, 15
    SQL pools, 15
        control nodes, 15
        DMS, 15
        logical architectural
            diagrams, 15–16
        MPP, 15
Synapse Studio, 15

T

tables
    partitions, Azure Synapse
        Analytics, 101–102
    replicated table distributions, 101
time window aggregations, stream-
    transport and processing engines,
    192
TinkerPop Gremlin API, 36
transformation, data streams, 190
triggers
    attaching to pipelines, 161–162
    containers, 41
    creating, 157–159
    event triggers, 161
    scheduling, 159–160
    tumbling window triggers,
        160–161
troubleshooting partitions, 260–263
TTL (Time To Live), 41
tumbling windows, 160–161, 203
type entry, ARM templates, 53

U

unstructured data, 3
updating connectors, 148
uploading data to storage, 59–60
    API, 66–67
    AzCopy, 65–66

Cosmos DB, 62–63
Microsoft.Azure.Storage.Blob, 68
Storage Explorer, 63–65
use cases, 26
    advanced analytics, 26
    hybrid ETL with existing
        in-premises SSIS, 26–27
    IoT architectures, 28
user names, Azure Data Studio,
    92–94

V

vCore model, Azure SQL database,
    14
vertical partitioning, 261
Visual Studio Code, changing ARM
    templates for storage, 54–56

W

watermarks, stream-transport and
    processing engines, 191
windows, Azure Stream Analytics,
    202–203
    hopping windows, 203–204
    session windows, 205–206
    sliding windows, 204–205
    tumbling windows, 203
workspaces, creating, 165–168

X

XML, semi-structured data, 2–3

Y - Z

ZRS (Zone-Redundant Storage),
    10, 83

Plug into learning at

MicrosoftPressStore.com

The Microsoft Press Store by Pearson offers:

- Free U.S. shipping

- Buy an eBook, get three formats – Includes PDF, EPUB, and MOBI to use with your computer, tablet, and mobile devices

- Print & eBook Best Value Packs

- eBook Deal of the Week – Save up to 50% on featured title

- Newsletter – Be the first to hear about new releases, announcements, special offers, and more

- Register your book – Find companion files, errata, and product updates, plus receive a special coupon* to save on your next purchase

 Pearson